The Cambridge Technical Series
General Editor: P. Abbott, B.A.

AN INTRODUCTION TO
APPLIED MECHANICS

AN INTRODUCTION

TO

APPLIED MECHANICS

BY

EWART S. ANDREWS, B.Sc. Eng. (Lond.)

Lecturer in the Engineering Department of the Goldsmiths' College,
New Cross, and of the Westminster Technical Institute
Formerly demonstrator and lecturer to the Mechanical Engineering
Department of University College, London
Member of the Council of the Concrete Institute

With numerous illustrations and numerical examples

Cambridge :

at the University Press

1915

CAMBRIDGE
UNIVERSITY PRESS

University Printing House, Cambridge CB2 8BS, United Kingdom

Cambridge University Press is part of the University of Cambridge.

It furthers the University's mission by disseminating knowledge in the pursuit of education, learning and research at the highest international levels of excellence.

www.cambridge.org
Information on this title: www.cambridge.org/9781316612804

© Cambridge University Press 1915

First published 1915
First paperback edition 2016

A catalogue record for this publication is available from the British Library

ISBN 978-1-316-61280-4 Paperback

PREFACE

MANY engineering and architectural teachers have found that applied mechanics is not an easy subject to teach, and most students have discovered that it is a difficult subject to understand. In searching for the reason for this unfortunate state of affairs, the author came to the conclusion that the treatment of the older form of text-book was too much that of applied mathematics—a kind of exercise-ground for algebraic manipulation—and that many of the more modern books that have attempted to remedy this weakness have given too much engineering application of the principles of mechanics without sufficient explanation of those principles.

The aim of the present book is to present the elementary principles of mechanics in accurate though clear terms and to show the application of those principles to the simpler problems arising in engineering and architectural applications. The general treatment is based more upon graphical conceptions than upon purely mathematical analysis because experience shows that the mind of the engineering student reasons more clearly from diagrams than from symbols.

A number of simple experiments have been given, principally those which require the simplest form of apparatus. It is not suggested that the experiments given are all that are desirable in a laboratory course, but it is believed that sufficient have been given to make the principles clear. It may be pointed out here that there is some danger in attempting to learn principles merely by experiments with simple (and usually inaccurate) apparatus. Before the student can hope to obtain valuable results from experiments, he must learn to make accurate readings of his instruments and to make corrections for the errors that may arise. Some authorities seem to suggest that experiment is of much greater importance to engineers than reasoning, but it should be borne in mind that training is required for good experimental work as well as for anything else, and in the author's opinion many engineering students who attempt to gather a knowledge of mechanical principles from experiment have not had sufficient preliminary training in experimental method. If our reasoning is based upon experimental laws and not upon dogmatic mathematical conceptions we shall probably make greater progress in elementary work by using experiment as an illustration of the results of our reasoning than by attempting to deduce the principles from the results of our experiments.

The great value of training in experimental work—and thorough training is essential—lies in the direction of research work which comes when we have understood the principles based upon the earlier researches of others.

It is hoped that this book will be found of value as a class-book in the junior classes of Engineering Colleges and in Public Schools that have an engineering side.

The author wishes to express his gratitude to Mr J. B. Peace, M.A., of Emmanuel College, Cambridge, for much valuable criticism and assistance with the proofs, and to the publishers for the great help that they have given in the preparation of the diagrams.

E. S. A.

GOLDSMITHS' COLLEGE,
NEW CROSS, S.E.
May 1915.

CONTENTS

CHAPTER I

FORCES AND OTHER VECTOR QUANTITIES

Diagrammatic representation of forces—Resultant of a system of forces; triangle of forces—Resolution of forces—Equilibrium; equilibrant—Vector polygon construction—Experimental errors . Pages 1-15

CHAPTER II

MOMENTS AND LEVERAGE

Positive and negative moments—The principle of moments—Reactions on a beam—Stability of a wall—Lever safety valve—Equilibrium of a body under three forces—Link and vector polygon construction—Couples 16-35

CHAPTER III

WORK, POWER AND ENERGY

Definitions—Kinetic and potential energy—Conservation of energy—Useful energy—Work done by a variable force—Work against resistance—Graphical representation of effort and resistance—Mean effort 36-51

CHAPTER IV

MACHINES AND EFFICIENCY

Wheel and axle and crow-bar—Mechanical advantage; efficiency of machines; velocity ratio—The inclined plane—The screw and screw-jack—Reversing machines—Pulley tackle—Weston's pulley block—Actual performance of machines—Indicated and brake horse-power 52-82

CHAPTER V

VELOCITY AND ACCELERATION

Uniform velocity—Velocity variable in magnitude—Velocity and space curves—Acceleration—Relation between acceleration, velocity and space curves—Equations of motion for constant acceleration—Gravity acceleration—Limits of use of simple formulae—Distance moved in a particular second 83-102

CONTENTS

CHAPTER VI

VELOCITY CHANGE IN DIRECTION; RELATIVE VELOCITY

Combination of velocities—Change of velocity—Relative velocity
103-112

CHAPTER VII

KINETIC ENERGY AND MOMENTUM

Measurement of kinetic energy—Connection between force and acceleration—Momentum—Importance of acceleration in traction problems
113-123

CHAPTER VIII

NEWTON'S LAWS OF MOTION; IMPACT

Newton's Laws—Impact and impulse—Equality of momentum before and after impact—Recoil of guns—Pile-drivers . . **124-138**

CHAPTER IX

STRESS AND STRAIN

Definitions—Hooke's Law—Stress-strain diagrams for mild steel, cast iron and concrete—Elastic moduli—Factor of safety—Resilience—Stress due to sudden loading—Temperature stresses . **139-160**

CHAPTER X

RIVETED JOINTS; THIN CYLINDERS

Forms of rivet heads and joints, and diameter of rivets—Methods in which a joint may fail—Efficiency of joint—Strength of thin cylinders and pipes **161-173**

CHAPTER XI

THE FORCES IN FRAMED STRUCTURES

Kinds of framed structures—Relation between bars and nodes in a perfect frame—Curved members—Reciprocal figure construction—Distinction between ties and struts—The method of moments . . **174-187**

CHAPTER XII

BEAMS AND GIRDERS

Shearing force and bending moment—Diagrams for standard cases of loading for cantilevers and simply supported beams—Graphical construction **188-202**

CHAPTER XIII

CENTRE OF GRAVITY AND CENTROID

Centre of gravity by moments—Centre of gravity as balance point and by inspection—Centroid of an area—Centroid of triangle, quadrilateral, trapezium, semicircle and parabola—Centre of gravity of pyramids and cones—Graphical construction for centroid—Kinds of equilibrium

203-223

CHAPTER XIV

FRICTION AND LUBRICATION

Static and kinetic friction—Coefficient of friction and angle of friction—Rolling friction—Inclined plane and screw with friction—Angle of repose—Efficiency of a screw—Lubrication .   .   .   **224-241**

CHAPTER XV

MOTION IN A CURVED PATH

Hodograph—Uniform motion in a circle—Centripetal and centrifugal force—Railway curves and motor tracks—Centrifugal governors—Balancing rotating parts—Projectiles **242-257**

CHAPTER XVI

MECHANISMS

Crank and connecting-rod mechanism—Instantaneous or virtual centre—Watt's parallel motion—Quick-return mechanism—Toggle mechanism —Cams and wipers—Pawl and ratchet mechanism .   **258-272**

CHAPTER XVII

BELT, CHAIN AND TOOTHED GEARING

Belt gearing—Velocity ratio—Speed-cones—Sizes of cones for keeping belt taut—Belt reversing gear—Belt drive for inclined axes—Toothed gearing—Rack, spur, bevel, spiral and worm gearing—Toothed gear trains—Idle gear wheels—Back gear for lathes—Reversing drive for lathe lead screw—Bevel gear reversing train .   .   .   **273-293**

APPENDIX

Sum curve construction **294**

Trigonometrical relations **297**

Mathematical tables **298**

Answers to Exercises **310**

Index **314**

CHAPTER I

FORCES AND OTHER VECTOR QUANTITIES

QUANTITIES which can be represented in magnitude and direction by straight lines are called *vector* quantities; the length of the straight line represents the magnitude of the quantity to some chosen scale and the direction of the straight line, as indicated by an arrow-head, represents the direction in which the quantity acts. The term *vector* is used in contra-distinction to the term *scalar*, scalar quantities being those which have magnitude only. Length is a good example of a scalar quantity; time is another. When we say that a body is 10 inches long we know everything about the length, but there are some quantities which are not fully defined until we know the direction as well. Forces and velocities are familiar examples of vector quantities; a vertical force of 10 lbs. is not the same thing as a horizontal force of 10 lbs. It should, however, be remembered that the direction and magnitude of a force does not tell us all that we wish to know about it; we must also know the *position* of the force, i.e., the actual position of the line of action of the force, because a force may be regarded as acting at any point in its line of action.

The scientific definition of a force is as follows: "*A force is that which alters, or tends to alter, the motion of a body in a straight line.*" This definition is based upon **Newton's** first law of motion which states that "a body continues in a state of rest or uniform motion in a straight line unless it be acted upon by some external force." It is not, however, essential to understand fully this definition of force at the present stage, because the idea of a force as a push or a pull is quite clearly understood by most people.

Weight. It is one of the fundamental laws of nature that between all bodies there exists a force of attraction and the earth exerts upon all bodies a force called the force of gravity tending to pull the body towards the centre of the earth. The weight of a body is the force exerted by gravity upon it and is the most familiar case of a force.

Unit of force. The weight of a given quantity of matter is found to vary with the latitude; it is about one-half per cent. greater at the poles than at the equator. We will take as our unit of force the weight of one pound in London; the pound being the quantity of matter in a standard cylinder of platinum preserved in London by the Board of Trade. This unit is often known as the "engineer's unit" or the "gravitation unit," to distinguish it from the "absolute unit" used by physicists. The absolute unit, which is the force required to produce a definite change of motion in a given mass in an assigned time, is independent of locality and is the more scientific of the two. As however engineers must be able to express their data and their results in the units in common use in their profession, and as the simultaneous use of two systems of units only leads to confusion, we shall confine ourselves to the engineer's unit as defined above.

Diagrammatic representation of forces. Referring to Fig. 1, suppose that a force F_1 acts through a point A in a body; as

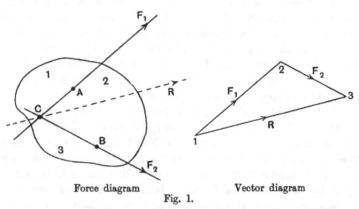

Force diagram   Vector diagram

Fig. 1.

previously noted, it is better to speak of a force as acting *through* a point than as *at* a point. If now in some convenient position we draw a line 1, 2 parallel to F_1 and of length to represent its magnitude to some convenient scale, 1, 2 will be the vector for the force F_1, the arrow-head representing the direction in which the vector is to be considered as acting. A very convenient method of indicating the force in many cases consists in numbering or lettering the space on each side of the force in the force diagram. The force is then denoted by the numbers or letters between which it acts; thus the force F_1 in the figure is called 1, 2, so that spaces on the force diagram correspond to points on the vector diagram. This is often referred to as " Bow's notation."

Resultant of a system of forces; triangle of forces. The resultant of a system of forces acting upon a body may be defined as the *single force which will have the same effect upon the body as the combined effect of the separate forces.*

Suppose that a second force F_2 acts through a point B in the body and let the lines of action of the two forces intersect at the point C. On the vector diagram, starting from the point 2, set out a length 2, 3 equal in length to the force F_2 to the scale already chosen for F_1 and parallel to F_2 in the direction of its arrow-head, and join 1, 3, the triangle 1, 2, 3 being commonly referred to as the *triangle of forces.* Then 1, 3 represents in magnitude and direction (but not in position) the resultant R of the two forces. It will act through the intersection C of the lines of action of the two forces, so that by drawing as shown in dotted lines a line R through C parallel to 1, 3 we can say that R is the resultant of the two forces F_1 and F_2 and that its magnitude is given by the length 1, 3.

It should be noted that the direction of the resultant is from the first point on the vector diagram to the last, i.e., from 1 to 3 ; by keeping this fact always in mind we shall avoid the confusion that sometimes arises. Further *the resultant does not act in the line* 1, 3 *but through the intersection C of the two forces.*

To make quite clear the plotting of the vector diagram suppose that F_1 is 12·5 lbs. and F_2 is 8 lbs. and that the vector diagram is drawn to a scale $1'' = 10$ lbs., then 1, 2 will be drawn 1·25 inches long and 2, 3 will be made ·8 inch long and if 1, 3 is 1·7 inches long, the resultant R of the two forces will be equal to $1·7 \times 10 = 17$ lbs. and will be in the direction shown. We need not restrict ourselves to actually measuring the resultant

R; we may calculate it when required by means of the trigonometrical methods of the solution of triangles. When we do calculate in this way it is not necessary to draw the triangle of forces accurately to scale. We do not propose to give a rigorous proof of this result at present but it may easily be verified experimentally in the following manner:

Experiment 1. Upon a suitable drawing board A (Fig. 2) arranged vertically, to which is fixed a piece of paper P, fix a spring balance by a string to a point B and connect a string to the hook of the balance and tie it to a ring D to which is tied another string connected to a weight F_2. Tie a third piece of string to the ring and pass the other end of it over a freely-mounted

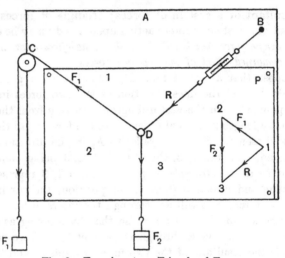

Fig. 2. Experiment on Triangle of Forces.

pulley C and connect a weight F_1 to its end. After the strings have come to rest, trace their directions upon the paper and remove the latter; then at some convenient place at the side draw to a suitable scale 1, 2 parallel to the portion DC of the string and of length to represent the weight F_1, and draw 2, 3 vertically and make it represent the weight F_2 to the same scale and join 1, 3. Then 1, 3 will be found to equal a length representing the reading upon the spring balance to the previously chosen scale and will be found parallel to BD. The pull in the portion CD of the string is equal to the weight F_1 on the end of it if there is perfect freedom of movement of the bearings of the pulley, so that our experiment will have verified the law of the Triangle of Forces. (This equality of tension on the two sides of the pulley follows from the Principle of Moments, p. 18.) Now suppose that one of the weights F_1 or F_2 is changed; the effect of this change will be that the strings will move and will finally come to rest in a new position; the directions and magnitudes of the forces will again be found to follow the triangle law.

Notation to represent vector addition. We have seen that
the resultant R is equivalent to the combined action of the
two forces F_1 and F_2. R is then said to be the *vector sum* of
F_1 and F_2. We may write this symbolically as

$$R = F_1 + F_2.$$

The +, which is a modified *plus* sign, indicates that the
addition is not a mere numerical or algebraic one, but that the
directions of the forces F_1, F_2 are taken into consideration in
effecting the addition.

Numerical Example—Thrust on a steam-engine foundation.
As a simple numerical example of the triangle of forces take

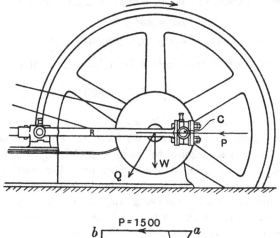

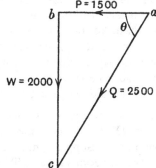

Fig. 3.   Thrust on Steam-Engine Foundation.

the horizontal steam-engine shown in Fig. 3. The force or
thrust upon the foundations, at the main bearing, for the position

shown of the crank-pin C and connecting-rod R, is made up of the weight W of the flywheel, shaft, etc., and the pull P exerted by the piston. Take $W = 2000$ lbs. and $P = 1500$ lbs. and set out ab horizontally to a convenient scale to represent P, say $1'' = 1000$ lbs. (1000 lbs. forms a very convenient unit for engineering calculations and has been called a " kip "); then set out bc vertically to the same scale to represent 2000 lbs. and join ac. Then ac gives the resultant thrust upon the foundations, a being the first point and c the last.

ac if scaled off will be found to be 2500 lbs. (2·5 kips), but we should note that we can find it by calculation in this case as easily as by scaling off, and consequently we need not draw accurately to scale, i.e., because abc is a right-angled triangle,

$$ac^2 = ab^2 + bc^2,$$

i.e. $ac = \sqrt{ab^2 + bc^2} = \sqrt{1500^2 + 2000^2} = \underline{2500 \text{ lbs.}}$

If we want to find the angle θ by trigonometrical calculation we note that

$$\tan\theta = \frac{cb}{ba} = \tfrac{2000}{1500} = 1\cdot3333,$$

and from trigonometrical tables we find that $\theta = 48° 35'$ nearly. It should again be noted that we have the choice of actually drawing the vector figure to scale and solving the problem graphically, or of using trigonometry or other mathematical means of calculation. The student should endeavour to be able to use both methods, each being appropriate in certain cases. If, however, at this stage he has no knowledge of trigonometry, this need not act as an obstacle to him; he can always use the graphical solution. We wish, however, to point out that many students fall into the mistake of never learning the mathematical method at all, and consequently waste time in many problems. A brief explanation of trigonometry is given in the appendix. In this case therefore we should write $2500 = 1500 + 2000$.

Resolution of forces. Let F, Fig. 4, represent a force acting through a point O and let OX, OY be two lines passing through O at any inclination whatever. Suppose that we wish to know the forces acting in these two directions which will have a resultant equal to F.

Set out a length 1, 2 to represent the force F: from one end, say 1, draw 1, 3 parallel to OX and from the other end draw

2, 3 parallel to OY, the intersection giving the point 3. Then
1, 3 represents the force F_X which acting in the direction OX will
combine with F_Y, represented by 3, 2, acting in the direction
OY, and have a resultant equal to F; this follows from the rule,
that we have already explained, that the resultant of any two
forces is given in magnitude and direction by the third side of
a triangle the other two sides of which represent the two forces
in magnitude and direction. The force F is then said to be
resolved into the forces F_X and F_Y, which are called the *components* of the force F in the two given directions.

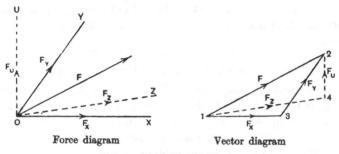

Force diagram   Vector diagram

Fig. 4.   Resolution of Forces.

In an exactly similar manner the force F could have been
resolved into components F_Z, F_U in the directions OZ, OU, as
shown in dotted lines intersecting at the point 4. We will
again emphasize the fact that it is not necessary to measure
the lengths 1, 3 and 2, 3; we may calculate them by trigonometrical or other methods whenever convenient. It is important
to note that *when components in two directions at right angles
are considered, the two components are called the resolved parts of
the force in the two directions and that a force has no "resolved part"
in a direction at right angles to its line of action.*

Numerical Example on resolution of forces. *A barge
A (Fig. 5) is being pulled along a canal by a horse which exerts a
force of* 150 *lbs. in a direction at* 20° *to the centre line of the canal.
Find the force urging the barge forward and that tending to pull
it into the side.* We require to find in this case the components
of the force of 150 lbs. in the direction AC and at right angles
to AC.

Draw a line ac parallel to AC, and drawing ab in a direction at 20° to ac to a scale say $1'' = 50$ lbs., i.e., making $ab = 3''$, draw bc perpendicular to ac. Then by measurement we should have

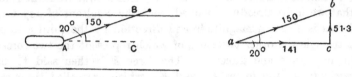

Fig. 5.

$bc = 1\cdot03'' = 51\cdot5$ lbs. = the force tending to pull the barge into the side and $ac = 2\cdot82'' = 141\cdot0$ lbs. = the force urging the barge forward along the centre of the canal. By calculation we should have

$$bc = 150 \sin 20° = 150 \times \cdot3420 = 51\cdot3 \text{ lbs.}$$
$$ac = 150 \cos 20° = 150 \times \cdot9397 = 141\cdot0 \text{ lbs.}$$

Equilibrium; equilibrant. If a body is at rest or is moving without altering its velocity, the forces acting upon it are said to be " in equilibrium." If such is the case, there is nothing tending to " alter its condition of rest or uniform motion " so that there is no resultant force acting upon it. We see therefore that the first essential of equilibrium of a body is that the resultant

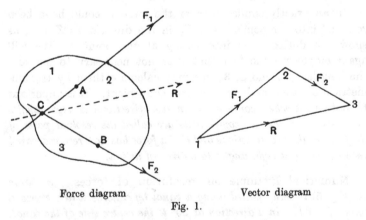

Force diagram   Vector diagram

Fig. 1.

of all the forces acting upon it shall be zero. This means that the first and last points of the vector diagram must coincide, because the distance between the first and last points gives us the value of the resultant.

Referring again to Fig. 1, we see that the body is not in equilibrium under the action of the forces F_1, F_2 because their resultant R is given by 1, 3 which is not zero. Now the single force that has to be added to a system of forces acting upon a body to bring the body into equilibrium is called the *equilibrant.* The *equilibrant must be equal and opposite to the resultant,* because the system may, as we have seen, be considered as replaced by the resultant and our forces then become reduced to two—the resultant and the equilibrant—and the resultant or combined effect of these two forces must be zero for equilibrium.   The only way for two forces to reduce to zero is for them to be equal and opposite; the only way for instance for the force R to be neutralised or equilibrated is for the equilibrant to be equal to 3, 1.

Warning; forces acting round a triangle. Care should be taken to distinguish between the following cases:

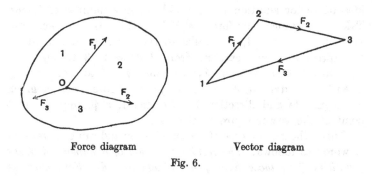

Force diagram   Vector diagram

Fig. 6.

(*a*) Three forces acting *at a point* and having their vector figure a closed triangle. This is the case that we have already considered and is shown diagrammatically in Fig. 6.

(*b*) Three forces acting on a body *round the sides of a triangle* are not in equilibrium even if the sides of the triangle are proportional to the forces. This is shown in Fig. 7. In this case the *three forces do* NOT *meet at a point* and the two forces F_1 and F_2 have a resultant *acting through B* which is equal to $F_1 + F_2 = F_3$ parallel to AC (if the forces are proportional to the sides of the triangle). The three forces are therefore equivalent to two equal and opposite parallel forces F_3 at a distance x apart. These two forces will tend to turn the triangle round in the direction

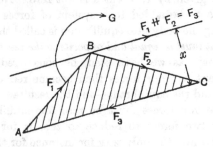

Fig. 7.

of the arrow G and form what is called a *couple* (which we shall consider more fully later).

More than two forces.   Vector polygon construction. Up to the present we have dealt with two forces only, but the idea of vector addition is applicable to any number of forces. Take for instance four forces $F_1F_2F_3F_4$ (Fig. 8): to some convenient scale draw 1, 2 parallel to represent F_1 in magnitude and direction; then starting from 2 draw 2, 3 to represent F_2; then 3, 4 to represent F_3; and finally 4, 5 to represent F_4. Then the resultant R of the whole system of forces will be given in magnitude and direction by the line 1, 5 joining the first point of the vector figure to the last.

This is the general case of vector addition and can be expressed in words as follows: *The resultant or sum of a number of vector quantities* (i.e., *those having magnitude and direction, such as forces and velocities or speeds*) *is obtained by placing them end to end, preserving their directions and a continuous sense of their arrow-heads; the final step from the beginning of the first vector to the end of the last is the resultant or vector sum.*

The reader may find this definition rather difficult to follow at first, but if he reads it carefully in connection with the figure the meaning should become clear. To express this result generally by a formula, where there are altogether n forces (where n is a whole number), the first of which is F_1 and the last F_n, we should write

$$R = F_1 + F_2 + \ldots . F_n.$$

This formula should be regarded merely as a symbolic or shorthand way of expressing the statement in italics.

Proof. This principle may be proved by repeating the construction for two forces, thus: It follows by the triangle of

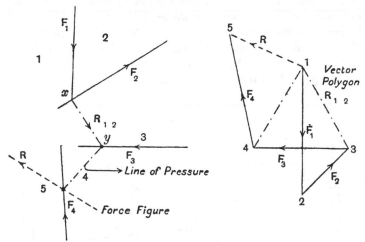

Fig. 8.   Vector Polygon construction.

forces that the length 1, 3 (shown in chain dotted lines) represents the resultant $R_{1,2}$ of the forces F_1, F_2; it must act through their intersection x as shown on the force figure; let the line of action of $R_{1,2}$ intersect the third force F_3 in y. Now 1, 4 represents the resultant of $R_{1,2}$ and F_3 and therefore of F_1, F_2 and F_3; draw therefore through y a line parallel to 1, 4 to intersect F_4 as shown. The force 1, 5 clearly represents the resultant of 1, 4 and F_4, i.e., of F_1, F_2, F_3 and F_4, and gives therefore the resultant R required and by drawing a line parallel to 1, 5 through the point last obtained we get the line of action of R.

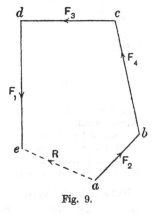

Fig. 9.

The line x, y, etc. is often called the *line of pressure* because one of its principal practical applications arises in problems relating to walls, arches and similar structures. We shall deal later with an extension of this graphical construction.

Order of taking the forces. It should be noted carefully that this construction gives the same result no matter what be the order in which the forces are taken, although the order will make an alteration in the shape of the vector polygon and for general convenience it is best as a rule to take them in turn. Fig. 9 for instance shows the vector polygon for the forces taken in the order F_2, F_4, F_3, F_1. The magnitude and direction of the resultant R is the same as before.

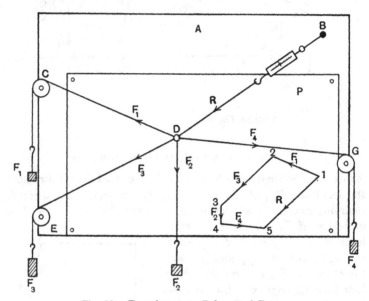

Fig. 10. Experiment on Polygon of Forces.

Experiment 2. This principle may be verified experimentally by an apparatus similar to that employed for the triangle of forces. Connect two additional pulleys E, G (Fig. 10) to the apparatus and pass strings over them, connected to the ring and carrying weights F_3, F_4 at their ends. Then by drawing the polygon of forces as shown we get the resultant R which should agree with the reading given by the spring balance.

Experimental errors. In this and all other experiments it should be remembered that it is difficult if not impossible to get absolute agreement between theory and experiment. This is due in part to the imperfections of our apparatus, introduced for example by friction in the pulleys and sagging of the strings due to their own weight. It is also caused by errors of observation; we shall probably not transfer the directions of the forces to our paper without slight errors and we may make some mistakes in drawing our parallels to get the polygon of forces. The existence of these experimental errors points to the absurdity of trying to express the results of calculations based upon experimental data by numbers carried to more than a few significant figures. If, for instance, we express the weight of a girder as 7·13762 tons we are laying claim to an accuracy which is impossible in practice. Manufacturers of steel plates, angle-bars, etc., cannot guarantee the sections nearer than about $2\frac{1}{2}$ per cent.; this means that if a girder is listed as weighing 100 lbs. per foot length, it may actually weigh anything between 97·5 and 102·5 lbs. per foot. The above-cited girder therefore should be put down as having a weight of 7·14 tons, the remaining figures being quite meaningless on account of the nature of the problem.

In this connection we would point out that *it is the number of significant figures that matters and not the number of figures after the decimal point.* If in the process of multiplication we get a result 581,574 lbs., we should write this as 582,000 lbs.; or if we had ·002876 foot we should call it ·00288 to the third significant figure, that is to the same degree of accuracy as in the previous case. The ordinary processes of long multiplication and division should be discarded in engineering calculations for logarithms or the slide-rule.

SUMMARY OF CHAPTER I.

The resultant of a system of forces acting upon a body is the single force which will have the same effect upon the body as the combined effect of the separate forces.

The *resultant* R of two forces F_1 and F_2 acts through the intersection of their lines of action and can be found by means of measurement or calculation from a triangle two sides of which are parallel to and proportional to the two forces, the direction of the forces being maintained. *This result is* written $R = F_1 + F_2$. A force can be *resolved* in any two directions by the aid of the triangle of forces, but a force has no resolved part in a direction at right angles to its line of action when the directions considered are at right angles to each other.

The *equilibrant* of a series of forces is equal and opposite to the *resultant*.

The principle of the triangle of forces can be extended to deal with any number of forces, the resulting construction being known as the " polygon of forces."

EXERCISES. I.

We give below a number of exercises for testing the extent to which the reader has followed the arguments so far. The reader will find by experience that he learns most thoroughly by working the examples as he proceeds and that it is better to do a little of the subject thoroughly than to press forward before each step is mastered.

1. Find the resultant of forces of 4 and 5 lbs. acting at 30° to each other.

2. A push of 36 lbs. acts horizontally at a point upon a roof-truss and at the same point inclined to it at an angle of 135° in an anti-clockwise direction is a pull of 70 lbs. Find the resultant force acting upon the truss at the given point.

3. Show that if the angle between two forces of given magnitude is increased, their resultant is decreased.

4. The greatest resultant that two given forces can have when acting in any direction is 100 lbs. and their least resultant is 20 lbs. Find their resultant when they act at right angles to each other.

5. The horizontal and vertical components of a certain force are equal to 5 and 12 lbs. respectively. What is the magnitude of the force?

6. A nail is being driven into a vertical wall at an inclination of 30° to the horizontal and a man pulls the nail horizontally away from the wall with a force of 10 pounds. Calculate the force tending to extract the nail and that tending to bend it.

7. A weight of 100 lbs. is suspended by wires from two points on a horizontal bar 5 feet apart, one wire being 6 feet long and the other 7 feet long. Find the forces in the two wires.

8. The thrust upon the horizontal foundation of a building is 100 tons inclined at 10° to the vertical. Find the force or pressure tending to drive the foundation into the ground and that tending to make it slide.

9. An inclined force of 200 lbs. is acting upon a body resting upon a horizontal surface, the horizontal resistance to movement of the body being 40 lbs. Find the smallest angle at which the force can act without moving the body.

10. A truck weighing 10 tons is resting against buffers in a siding the slope of which is 1 in 30; what is the pressure on the buffers ?

11. Find the resultant of two forces of 15 and 36 lbs. acting at an angle of 52° with each other.

12. A weight of 75 lbs. is carried by two cords which make angles of 35° and 51° with the horizontal. Find the pull in each cord.

13. Forces $OA = 30$ lbs., $OB = 50$ lbs., $CO = 15$ lbs., $DO = 80$ lbs., $OE = 150$ lbs. meet at a point: the angles are $BOA = 45°$, $COA = 90°, DOA = 135°, EOA = 270°$. Find the resultant.

14. Let P_1, P_2, P_3 be three forces acting at O each $= 100$ lbs. And let $\angle P_1OP_3 = 135°$ and $\angle P_1OP_2 = 90°$. Find their resultant. (See Fig. I a.)

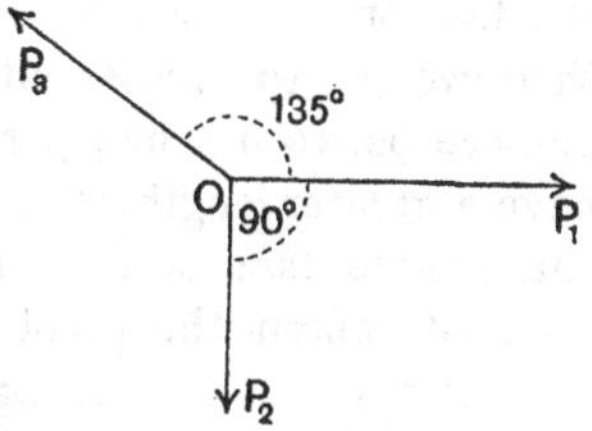

15. A ball weighing 100 lbs. is suspended from the ceiling by a string 8 ft. long. Find the force necessary to hold the weight 2 ft. from the vertical by a horizontal pull.

Fig. I a.

CHAPTER II

MOMENTS AND LEVERAGE

We have considered forces up to the present only from the point of view of their magnitude and direction; we will now consider them from another point of view which is extremely useful in engineering problems, viz., by their *moments* or *leverage**. Children accustomed to playing on a " see-saw " are aware of the fact that by sitting farther away from the pivot they use their weight to greater advantage, and that in order to get a balance between a heavy and a light child, the light child must have a greater length of plank. The scientific way of expressing this simple fact is that for the two forces to balance their moments about the pivot must be equal.

The moment or leverage of a force about any point may be defined as " *the tendency of the force to rotate the body, upon which it acts, about the point.*" It is measured by the product of the force into the perpendicular distance from the point to the line of action of the force. Referring to Fig. 11, the moment of the force F_1 about O is equal to $F_1 \times p_1$. If therefore F_1 is 15 lbs. and p_1 is 3 inches, the moment of F_1 about $O = M_1 = 15 \times 3 = 45$ pound-inches (or inch-pounds, but the former is preferable for a reason that we will explain later). The perpendicular distance from the point to the line of action of the force may be called the " arm " so that we get the general rule Moment = Force × arm.

Positive and negative moments. Since rotation can be in one of two opposite directions, we must distinguish between these two directions by calling one positive and the other negative. The tendency of the force F_1 is to rotate the body about O in the direction of the hands of a clock, i.e., a clockwise direction, and will for convenience be called a positive moment. If the

---

* The term "leverage" is often used in a more restricted sense than the above to denote the "mechanical advantage" (p. 53). We think, however, that it is better to use it as synonymous with "moment."

tendency of the force were to cause rotation in the opposite or anti-clockwise direction, as is the case with the force F_2, we should call the moment negative. It is purely a matter of convenience as to which is called positive and which negative. All that matters is that we shall agree to call positive the tendency to turn in one direction, and to call negative the tendency to rotate in the opposite direction.

Suppose for instance that $F_2 = 20$ lbs. and that p_2 is 1·6 inches, then the moment of F_2 about $O = M_2 = -20 \times 1 \cdot 6 = -32$ pound-inches.

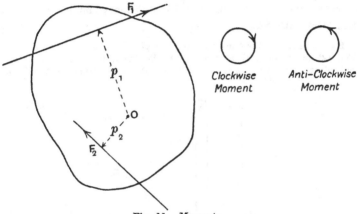

Clockwise Moment

Anti-Clockwise Moment

Fig. 11. Moments.

Now so long as we are dealing with forces in one plane moments are *scalar* quantities; they have no direction in the sense that a force or a velocity has direction and are added by the ordinary or algebraic rules.

Thus the total moment of F_1 and F_2 about $O = 45 - 32 = 13$ pound-inches.

Moment about a point in the line of action of a force. Remembering that a force must be considered as acting in a line rather than at any point, we shall see at once that a force ha's zero moment about a point in its line of action because its arm is zero so that the product of the force by the arm must also be zero.

The Principle of Moments. This principle, which is of very great value in engineering problems and a clear understanding of which will obviate many difficulties that might otherwise arise, may be stated as follows: *If a system of forces in one plane act upon a body and keep it in equilibrium, the algebraic sum of their moments about* ANY *point in the plane will be zero.*

By algebraic sum we mean the sum allowing for some being positive and some negative, so that we might say that the total positive moment must equal the total negative moment. We will not give a rigorous proof of this principle but will point out that it really follows from the idea of equilibrium and from Newton's first law of motion (p. 124). If a body is in equilibrium, it does not tend to change its state of rest so that there is no tendency to rotate about any point. We will restrict our consideration for the present to stationary bodies and will deal later with rotating bodies.

Experiment 3. *Verification of the principle of moments.* Take a rod *C* (Fig. 12) and pivot it about one end *A*, allowing it to hang freely. At a convenient point *B* attach a string to the rod and pass the string over a pulley *E*,

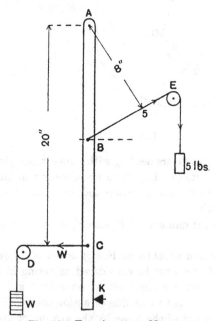

Fig. 12. Experiment on Moments.

at a point for instance above *B*, and hang a weight say of 5 lbs. at the end of the string. Provide a convenient stop *K* to prevent the pull from moving the rod out of position. At a convenient point *C* attach another string and

pass it over a second pulley D and weight the end of the string carefully until
the rod just begins to come away from the stop. So far as rotation about
A is concerned the rod is then in equilibrium under the two forces 5 and W,
there being now no pressure or " reaction " on the stop. Now measure carefully the perpendicular distances from A to the two strings BE and CD; it
should be noted that the distances are measured to the strings which represent
the lines of action of the forces and not to the points B, C at which the strings
are attached to the rod.

Then taking moments about A we have

$$M_A = W \times 20 - 5 \times 8 = 0 \text{ (because the rod is in equilibrium),}$$

$$\therefore 20W = 40 \text{ or } W = \tfrac{40}{20} = 2 \text{ lbs.}$$

W will be found to be approximately 2 lbs. in the experiment. It may not
be *exactly* 2 lbs. because, as we have already indicated, there are always
slight experimental errors especially with rough apparatus.

*The sum of the moments of a system of forces about a given
point is equal to the moment of the resultant about the same point.*

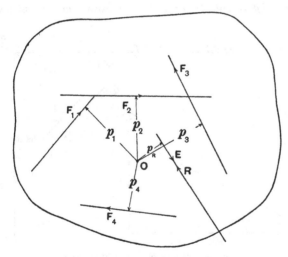

Fig. 13. The Principle of Moments.

This rule can be deduced from the general principle of moments
as follows. Let F_1, F_2, F_3, etc. (Fig. 13) be any number of forces
acting upon a body (in the figure we have shown four forces
but there may be any number) and let their resultant be R.
Now suppose that we add a force E, the equilibrant, which as
we have shown already is equal and opposite to the resultant.

Then the forces F_1, F_2, etc. and E keep the body in equilibrium, so that the total moment about any point O must be zero:

i.e. $F_1 p_1 + F_2 p_2 - F_3 p_3 + E p_R + F_4 p_4 = 0$,

$\therefore F_1 p_1 + F_2 p_2 - F_3 p_3 + F_4 p_4 =$ sum of moments of given forces

$$= - E p_R.$$

But $E = - R$,

$\therefore F_1 p_1 + F_2 p_2 - F_3 p_3 + F_4 p_4 = - (- R p_R) = R p_R$;

i.e. The sum of the moments of the system = Moment of resultant.

To save writing a long string of similar quantities, it is usual to use the Greek letter Σ (" sigma ") to indicate " sum of quantities like," so that in our present case we would write

$$\Sigma (F_1 p_1) = R p_R.$$

We shall find this notation very useful in other problems.

   Numerical verification. *Take the weighted lever SQ (Fig. 14), which is weighted with 10 lbs. at Q, and pivoted at a point 8 inches*

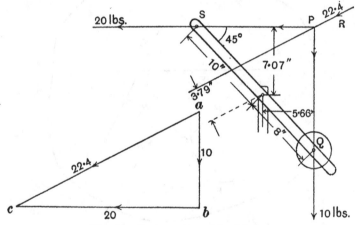

Fig. 14. Forces on a weighted lever.

from Q and is subjected to a horizontal pull of 20 lbs. at the point S the length SQ being 18 inches. The point of support or pivot of a lever is called the *fulcrum*. If the weight of the lever itself may be neglected, the forces tending to turn the lever about the fulcrum are the horizontal force of 20 lbs. acting through S and the vertical force of 10 lbs. due to the weight acting through Q. The lines of action of these two forces intersect at the point P;

by means of the triangle of forces we find the resultant ac of the forces of 10 and 20 lbs., which by measurement or calculation comes to 22·4 lbs. (R). Draw through P a line parallel to this resultant.

The perpendicular distance from the fulcrum to this line will be found by measurement to be 3·79 inches. So that the moment of the resultant about the fulcrum will be $22·4 \times 3·79 = -84·8$ pound-inches.

Now take moments of the separate forces about the fulcrum. These are equal to $-20 \times 7·07 + 10 \times 5·66 = -84·8$ pound-inches as before.

APPLICATIONS OF THE PRINCIPLE OF MOMENTS

We will now consider some applications of the principle of moments to practical problems. Further applications will arise at later portions of the book.

Reactions on a beam. *Determine the reactions on a beam of 20 feet span loaded in the manner shown in Fig. 15.* By " reactions " on a beam we mean the pressure exerted by the support

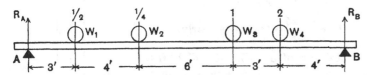

Fig. 15. Reactions on a beam.

upon the beam. If the support is stationary, it will press upward upon the beam with a force equal to that with which the beam presses against the support. This is an example of *Newton's Third Law of Motion* that " Reaction is equal to action." This " reaction " may be regarded as a force induced to counteract the original force (or " action ") so as to bring the resultant force at the point to zero.

In our present example there are six forces acting on the beam which keep it in equilibrium, viz., the four weights W_1, W_2, W_3 and W_4 and the two reactions R_A and R_B.

In the present case our forces are all vertical, the weights acting downwards and the reactions upwards. We have seen that, for equilibrium, the vector sum must be zero. When the forces are all in the same direction, the vector sum is the same as the algebraic sum so that we get the rule that "the total upward force must be equal to the total downward force."

Therefore we have:

$$R_A + R_B = W_1 + W_2 + W_3 + W_4 = \tfrac{1}{2} + \tfrac{1}{4} + 1 + 2 = 3\tfrac{3}{4} \text{ tons.}$$

To determine one of the reactions, say R_A, take moments about the other support B. The moment of R_B about B is zero because the "arm" is zero.

The algebraic sum of all the moments about B must be zero, i.e.

$$R_A \times 20 - W_1 \times 17 - W_2 \times 13 - W_3 \times 7 - W_4 \times 4 = 0,$$

i.e. $20 R_A = 17 \times \tfrac{1}{2} + 13 \times \tfrac{1}{4} + 1 \times 7 + 2 \times 4$

$$= 26 \cdot 75 \text{ tons-feet.}$$

$$\therefore R_A = \frac{26 \cdot 75}{20} = \underline{1 \cdot 34 \text{ tons nearly}}$$

$$\therefore R_B = 3 \cdot 75 - 1 \cdot 34 = \underline{2 \cdot 41 \text{ tons.}}$$

As a check on the working we will now find R_B by taking moments about A. We then have

$$20 R_B = \tfrac{1}{2} \times 3 + \tfrac{1}{4} \times 7 + 1 \times 13 + 2 \times 16$$

$$= 48 \cdot 25 \text{ tons-feet.}$$

$$\therefore R_B = \frac{48 \cdot 25}{20} = 2 \cdot 41 \text{ tons nearly.}$$

Stability of a wall. *A wall 18 ins. thick and 8 feet high weighs 6 tons. Find what horizontal pressure, due to the wind acting at the centre of the wall, would be necessary to overturn the wall.* Referring to Fig. 16, the forces acting upon the wall are the horizontal wind pressure P and the weight W which may be taken as acting down the centre line of the wall.

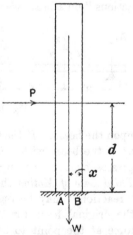

Fig. 16.   Stability of a wall.

Taking moments about the point B we have a clockwise moment of $P \times d$ called the "overturning moment" and an anti-clockwise moment of $W \times x$ called the "stability moment." If the overturning moment is less than the stability moment, the wall will not overturn, but if the overturning moment is ever so little greater

than the stability moment the wall will topple over. In the
limiting case in which the two moments are equal, the wall is
just about to overturn, and the value of P when the moments
are equal is usually taken as the least value required to cause
overturning. In this case therefore we have

$$P \times d = W \times x,$$

or $$P = \frac{Wx}{d}.$$

Now $W = 10$ tons, $d = 4$ ft. and $x = \frac{9}{12}$ ft.

$$\therefore P = \frac{10 \times 9}{4 \times 12} = \underline{1.875 \text{ tons.}}$$

The Lever Safety Valve. The lever safety valve provides
another common example of a device in which the principle

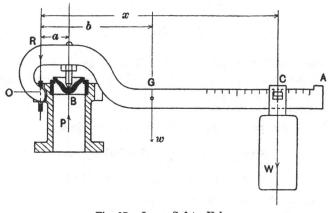

Fig. 17.   Lever Safety Valve.

of moments is used to facilitate calculations.   The device consists
of a lever AO (Fig. 17), pivoted* at O, and provided with a weight
W the position C of which is capable of adjustment along the lever.
A force P due to the pressure of steam in the boiler acts at the
point B and the parts are so proportioned that when the pressure
of steam in the boiler reaches a pre-determined limiting value of
safety, the force P is sufficient to lift the lever and allow the
steam to escape until the pressure falls to the required extent.
    There are four forces acting upon the lever: the upward

---

* The point about which a lever is pivoted is commonly referred to as the
"fulcrum."

force P acting through the point B; the movable weight W acting downwards through the point C; the weight w of the lever itself acting through a point G called the " centre of gravity," the position of which is found in the manner which we shall describe later; and finally the downward reaction R caused by the lever pressing upwards upon its pivot or fulcrum. In calculations we want to know the pressure P required to lift the valve for a given position of the weight W upon the lever.

By taking moments about the fulcrum O we get rid of the force R because its moment about a point in its line of action is zero. We will show later that we might have taken moments about any other point but that it would take longer.

Then we have: Clockwise moment about $O = Wx + wb$;

Anti-clockwise moment about $O = Pa$;

$$\therefore Pa = Wx + wb \quad \dots\dots\dots\dots\dots(1),$$

or
$$P = \frac{Wx + wb}{a} \quad \dots\dots\dots\dots\dots(2).$$

Numerical Example. *Take* $W = 62$ *lbs.,* $w = 12$ *lbs.,* $x = 33$ *ins.,* $b = 16\frac{1}{2}$ *ins.,* $a = 3\frac{1}{4}$ *ins. If the diameter of the valve be* 3 *ins. find the pressure in the boiler at which the safety valve will " blow off."*

Let p lbs. per sq. in. be the pressure required.

$$\text{Area of valve} = \frac{\pi}{4} \times 3^2 = 7\cdot07 \text{ sq. ins.}$$

$\therefore$ Total pressure $P = p \times \text{area} = 7\cdot07p$.

Putting the values in (2) we have

$$7\cdot07p = \frac{62 \times 33 + 12 \times 16\cdot5}{3\cdot25} = \frac{2244}{3\cdot25};$$

$$\therefore p = \frac{2244}{7\cdot07 \times 3\cdot25} = \underline{97\cdot7 \text{ lbs. per sq. in.}}$$

Calibration of Safety Valve. By "calibration" of an instrument is meant the determination of the scale for measuring the quantities with which the instrument deals. Suppose that we wish to mark points along the lever of the safety valve which we considered in the previous example to correspond to pressures from 60 to 100 lbs. per sq. in., rising by 5 lbs. per sq. in. at a time. Suppose that when the pressure is p lbs. per sq. in. the lever is in equilibrium with the weight w at a distance x from the point O. We have seen that $P = 7{\cdot}07p$.

Therefore taking moments as in equation (1) we have

$$7{\cdot}07p \times 3{\cdot}25 = 62x + 12 \times 16{\cdot}5,$$
$$\therefore 23p = 62x + 198,$$
$$\therefore 23p - 198 = 62x,$$
$$\therefore x = ({\cdot}371p - 3{\cdot}19) \text{ inches} \quad \ldots (3).$$

Equation (3) is what is called a "linear" equation because if values of the lever-length x be plotted against the pressure p the resulting diagram or graph will be a straight line.

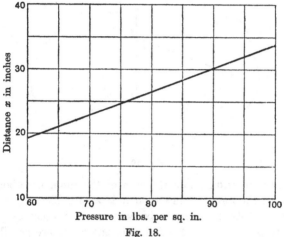

Fig. 18.

Fig. 18 shows the graph which is the "calibration curve" of the safety valve.

For $p = 60$ lbs. per sq. in., $x = {\cdot}371 \times 60 - 3{\cdot}19 = 19{\cdot}1$ inches.

For $p = 100$, we get $x = 37{\cdot}1 - 3{\cdot}19 = 33{\cdot}9$ inches.

The distance therefore between the 100 and the 60 marks on the lever is $33{\cdot}9 - 19{\cdot}1 = 14{\cdot}8$ inches and we have seen from the diagram that the divisions are equally spaced. Therefore the length of each of the eight divisions corresponding to 5 lbs. per sq. in. will be $\dfrac{14{\cdot}8}{8} = 1{\cdot}85$ inches.

Equilibrium of a body under three forces. There are a number of problems in which the number of forces acting can be

reduced to three. We can then make use of the following rule.
*If three forces act upon a body and keep it in equilibrium, they
must be in one plane* and are either parallel or their lines of action
meet at a common point.* We will first prove the rule by assuming that it is not true and that three forces F_1, F_2, F_3 (Fig. 19)
act upon a body and keep it in equilibrium. Let the lines of
action of F_1 and F_2 meet at A and suppose that the line of action
of F_3 does not pass through A. Now take moments about A.
The moments of F_1 and F_2 about A are each zero, but F_3 has a
moment of $F_3 \times p$ so that the sum of the moments of the three
forces about A is equal to $F_3 \times p$; we have seen, however, that

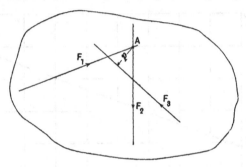

Fig. 19.

if a body is in equilibrium the sum of the moments about *any
point* of all the forces acting upon it is zero. The only possible
way for the total moment to be zero in the present case is for
F_3 also to pass through A. If two of the forces are parallel, the
third must also be parallel to them; if not the third force will
intersect one of the others at some point and there would be a
resultant moment about that point due to the third force. The
student should draw his own diagram to illustrate this.

This fact of the concurrency of three forces which keep a
body in equilibrium is employed in several problems involving the
reactions at the supports of structures. As a simple illustration
take the case of a lever AB (Fig. 20) pivoted at the lower point B
and held in the inclined position shown by a horizontal force P
acting through the point A. The three forces acting upon the
lever are the weight W acting vertically through the point G,

---

* The general consideration of forces not in one plane lies outside the scope
of this book. One simple case is dealt with on p. 184.

the horizontal pull P and the reaction R at the pivot at B. The
weight and pull act in given lines which intersect at the point O;
the reaction R must therefore also pass through O, so that by
joining OB we get the *line of action* of the reaction R. We have
now the directions of three forces in equilibrium, but have still
to determine their magnitudes. This is effected by drawing
the triangle of forces as already explained, setting down 1, 2 to

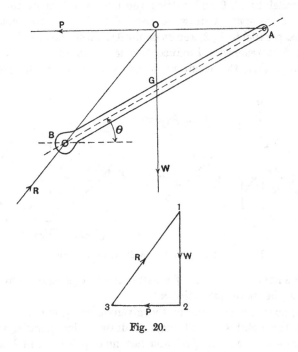

Fig. 20.

represent the weight W to convenient scale; through 2 draw a line
parallel to the force P and through 1 draw a force parallel to the
reaction R. The intersection 3 gives the third point required
in the triangle, and 2, 3 will represent the force P and 3, 1 the
force R to the same scale as that to which 1, 2 represents W.

**Graphical construction for moments; link and vector
polygon construction.** We have dealt already (p. 10) with the
graphical construction for finding the magnitude and direction
of the resultant of a number of forces. We now come to an
extension of that construction.

Let 0, 1; 1, 2 (Fig. 21), and so on, be a number of forces not necessarily parallel nor concurrent. To some suitable scale set down on a vector figure 0, 1, 2, and so on, then as before, the closing line 0, 5 gives the magnitude and direction of the resultant. Now take any point or pole P at any convenient position on the paper and join P, 0; P, 1; and so on. Then draw anywhere across the line of action of the first force a line af parallel to P, 0 and cutting the line of action of the force in a; across space 1 draw ab parallel to P, 1; across space 2, draw cd parallel to P, 3, and so on until the last line or link parallel to P, 5 is reached. Produce this last link to meet the first link in f, then the resultant R will pass through the point f,

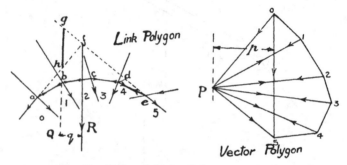

Fig. 21.   Link and Vector Polygon construction.

and the figure a, b, c, d, e, f is called the *link polygon* or by some writers the *funicular polygon*.

Suppose the moment of the given force system is required about the point Q.   Through Q draw a line parallel to the resultant R to cut the first and last links produced in h and g. Then if the point P is at perpendicular or polar distance p from 0, 5 on the vector figure, the moment of force system about Q is equal to $gh \times p$, gh being read on the space scale and p on the force scale.

Proof. By the law of vector addition, the force 0, 1 on the vector figure is equivalent to forces $0P$, $P1$ acting in fa and ab; the force 1, 2 is equivalent to forces $1P$, $P2$ acting in ba and bc, and so on, the last force 4, 5 being equivalent to forces $4P$, $P5$ acting in de and fe. It will be seen that with the exception of the forces down fa and fe all these forces neutralise each other, and so the resultant of the whole system of forces is the same as that of *fa* and *fe* and therefore acts at the point of intersection *f* of these forces.

The triangles *fgh*, P, 0, 5, are similar, the corresponding sides being parallel.

$$\therefore \frac{gh}{q} = \frac{0,5}{p}$$

(because in similar triangles the bases are proportional to the heights),

$$\therefore p \times gh = 0, 5 \times q,$$

but 0, 5 = resultant *R* and *q* is distance of *R* from *Q*,

$$\therefore 0, 5 \times q = \text{moment of force system about } Q,$$
$$\therefore p \times gh = \text{moment of force system about } Q.$$

Numerical Example. *Find the resultant of the loads shown in Fig. 22 and find their moment about the point A.* This is the

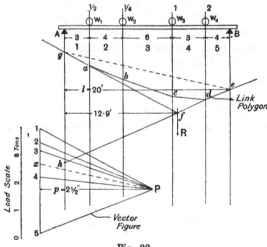

Fig. 22.

same system of loading as we considered analytically in the example on p. 21.

Number the spaces 1, 2, 3, etc. between the forces, and choosing a convenient scale of loads set down the vector figure 1, 2, 3 .. 5, which is a vertical straight line in this case because all the forces are vertical. Then, choosing any convenient pole P, join P to 1, 2 .. 5, draw anywhere across the first force line a line *af* parallel to *P*1, cutting it in *a*; then across space 2 draw *ab* parallel to *P*2; across space 3 draw *bc* parallel to *P*3; across space 4 draw *cd* parallel to *P*4; and through *d* draw *df* parallel to *P*5 cutting the first link *af* in *f*. Then the resultant *R*, which is equal to 1, 5 on the vector figure, i.e., 3·75 tons, acts through *f* which is at a distance 12·9 ft. from the left-hand end *A* of the beam. To find the moment of the system of forces about *A* draw a vertical through *A* and let *fa* produced meet it in *g* and *df* produced meet it in *h*. Then, according to the construction proved above, *gh* × *p*, i.e., 19·3 × 2·5 = 48·25 tons-ft., is the moment of the system of forces about *A*.

Extension of construction to find reactions. If we want to find the upward reaction of R_B at the support *B* we could then divide as before this moment by the " arm " of the reaction, i.e., by 20. We can, however, extend the construction as follows to do this division graphically. Let the last link *fd*, produced, cut the vertical through the support *B* at a point *e*. Join *ge*, as shown in dotted lines, and through the pole *P* draw a line *Px* parallel to *ge* to cut 1, 5 in *x*, then 5*x* will equal the reaction R_B and *x*1 will equal the reaction R_A (because as we have seen already the sum of the reaction must be equal to the total load if the beam is in equilibrium).

Proof. The triangles *Px*5 and *egh* are similar because their corresponding sides are parallel.

∴ since their bases are proportional to their heights we have

$$\frac{x5}{p} = \frac{gh}{l};$$

$$\therefore x5 = \frac{p \times gh}{l}$$

$$= \frac{\text{moment of force system about } A}{l}.$$

But we have by the principle of moments that

$$R_B \times l = \text{moment of force system about } A,$$

$$\text{i.e., } R_B = \frac{\text{moment of force system about } A}{l};$$

$$\therefore x, 5 = R_B.$$

Alternative proof. We can prove this in a similar manner to that on p. 28 as follows: $W_1 = 1, 2$ can be replaced by its components $1P$ in ga and $2P$ in ba; $W_2 = 2, 3$ by its components $2P$ in ab and $P3$ in cb and so on. Then $P2$ in ba balances $2P$ in ab and so on, so that we are left with only $1P$ in ga and $P5$ in ed. These meet in f so that the resultant R passes through f.

Also $1P$ in ga can be replaced by components xP in ge and $1x$ in Ag, and $P5$ can be replaced by Px in eg and $x5$ in Be. The forces in ge balance and the system reduces to $1x$ down at A and $x5$ down at B and these forces are equal and opposite to the reactions at A and B.

We will deal further with this construction later in considering calculations for beams and girders.

Couples. When the forces acting upon a body reduce to two equal and opposite parallel forces, they are said to form a *couple*. Thus the forces F (Fig. 23) form a couple. A couple

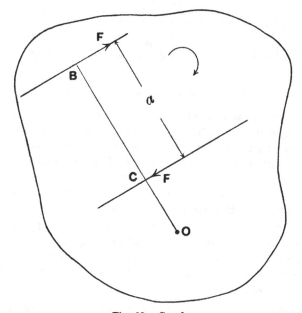

Fig. 23. Couples.

has no resultant because the vector sum of the forces composing it is zero (i.e., $F - F = 0$); but about any point O in the plane of the forces the couple has a moment equal to $F . a$ (a is the perpendicular distance between the forces and is called the *arm*).

To prove this take moments about the point O. Then we have

$$M_0 = F \times BO - F \times CO = F(BO - CO) = F . BC = F . a.$$

The effect therefore of this couple will be to rotate the body upon which it acts, without moving the body as a whole. The couple shown is clockwise.

The only way in which a couple can be neutralised or equilibrated is by the introduction of another couple of equal moment but opposite direction. The vector sum of the forces will still be zero and the moment about any point will also be zero, both conditions of equilibrium being therefore satisfied.

We have an example of couples in the tool shown in Fig. 24 diagrammatically, for enlarging holes in wood. The tool has

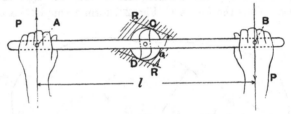

Fig. 24. Couple acting upon cutting tool.

two cutting lips and is operated by a lever AB which is grasped by the operator's hands. At C and D, the points of contact of the cutting lips with the wood, resisting forces R are brought into play tending to prevent the tool from rotating.

The operator exerts a couple of moment $P \times l$ and the resistance exerts a couple of moment $R \times a$. When the operator starts to press on the lever, it does not move; it is then in equilibrium and $P \times l$ the moment of the operative couple is equal to $R \times a$ the moment of the resisting couple. The operator then presses more strongly until the tool moves round, i.e., until he has exerted a couple greater than the resistance of the wood can exert.

Conditions of equilibrium in link and vector polygon construction. We have seen for a system of forces to be in equilibrium the conditions to be satisfied are (a) that the resultant is zero, (b) that the total moment of all the forces about any point must be zero. In the graphical construction, (a) is satisfied if the first and last points of the vector polygon coincide, i.e., if the vector polygon closes. To satisfy condition (b) the distance gh in Fig. 21 must be zero for every position of the point Q, in other words g and h must coincide. The only way in which this can happen is for the first and last links to coincide, or for the link polygon to close. If the vector polygon closes but the link polygon does not close, the system reduces to a couple. Since, as we have seen, a couple has the same moment about any point, we should expect that the first and last links must be parallel. If we consider the construction we see that this must be the case because the first and last points 0, 5 of the vector polygon coincide. Therefore the first and last links will both be drawn parallel to P, 0 and must therefore be parallel to each other.

SUMMARY OF CHAPTER II.

The moment of a force about any point is measured by the product of the force by the perpendicular distance from the point to the line of action of the force.

If a system of forces in one plane act upon a body and keep it in equilibrium, the algebraic sum of their moments about *any* point in the plane will be zero.

The sum of the moments of a system of forces about a given point is equal to the moment of the resultant about the same point.

If three forces act upon a body and keep it in equilibrium, they are in one plane and are either parallel or their lines of action meet at a common point.

When two equal forces are parallel and opposite in direction they are said to form a *couple*. The moment of the couple is measured by the product of one of the forces by the *perpendicular* distance between them. And a couple can be neutralised only by the introduction of another couple of equal moment but opposite direction.

The link and vector polygon construction enables us to determine graphically the position, direction and magnitude of the resultant of a number of forces and the moment of the resultant about any point.

If the system is in equilibrium, the link and vector polygons are both closed; if the system reduces to a couple the vector polygon is closed but the first and last links of the link polygon are parallel.

EXERCISES. II.

1. Define the moment of a force. A lever AB is hinged at A and carries weights as shown (Fig. II a). What force P acting upwards will keep the bar in a horizontal position?

2. A lever AB whose weight is 120 lbs. and length 3 feet has a fulcrum 10 inches from the end A. What weight at the end B will balance 384 lbs. placed at A? Also find the pressure on the fulcrum.

3. A uniform lever 26 inches long and weighing 45 lbs. carries a weight of 20 lbs. at one end and 35 lbs. at the other. Find the point in the lever about which it will balance.

4. A uniform lever 8 ft. long weighs 42 lbs. It carries a weight of 36 lbs. at one end and 24 lbs. at the other. Find the point about which it will balance.

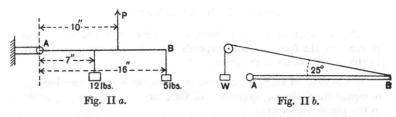

Fig. II a.   Fig. II b.

5. A safety valve is 3 inches in diameter and the weight on the end of the lever is 55 lbs., the distance of the fulcrum from the centre of the valve being 4·5 inches. If the weight of the lever and valve are negligible, how far along the lever from the centre of the valve must the weight be placed if the valve is to blow off at a pressure of 80 lbs. per sq. inch?

6. The area of a safety valve is 8 square inches. The lever is 2 ft. 6 ins. long, its centre of gravity is 1 ft. from the fulcrum and its weight is 10 lbs. The fulcrum is 4 inches from centre line of valve. Find the pressure in lbs. wt. per sq. inch at which steam will blow off, if the weight on end of lever is 65 lbs. and the valve itself weighs $1\frac{1}{2}$ lbs.

7. A lever 16 inches long weighs 25 lbs. and has a fulcrum at one end. It is held in a horizontal position by a vertical force applied at the other end. The lever being uniform, what is the magnitude of this force?

8. A uniform lever AB whose weight of 15 lbs. acts at the centre is 15 inches long; it is hinged at A and held horizontally by a cord carrying a weight W as shown (see Fig. II b). Find the magnitude of W.

9. A pole AC pivoted at C and carrying a weight of 1 ton is supported by a rope AB. Prove that the pull in AB will be least when its direction is at right angles to AC. Find this pull, and the thrust in the pole. (See Fig. II c.)

10. A system of five parallel forces whose magnitudes are 10, 12, 8, 6, 11 lbs. weight respectively act in lines 2 ins. apart. Find the position of their resultant.

11. A bent lever ACB is pivoted at C; the arm AC is horizontal and 9 inches long; the arm BC is vertical and 39 inches long. A load of 300 lbs. is hung from A. Find what horizontal force at B will produce equilibrium, neglecting the weight of the lever.

12. A beam 20 ft. long supported at its ends has a load of 2 tons at the centre of the span, another of 1 ton at 3 ft. from one end, and another of 3 tons at 4 ft. from the other end. Find the reactions of the supports neglecting the weight of the beam.

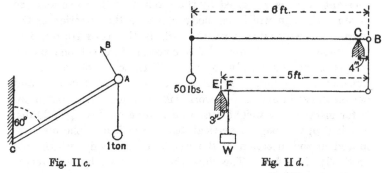

Fig. II c.   Fig. II d.

13. Fig. II d shows a compound lever. The fulcra are at C and E. Find the weight W which can be supported by an effort of 50 lbs. applied as shown, neglecting the weight of the beams.

14. Forces of 1, 2 and 3 lbs. are parallel and act at the corners of an equilateral triangle. Find where the resultant acts.

15. In an 8-oar boat each man pulls with a force of 60 lbs. If the oars are 10 ft. long and 2 ft. 6 ins. from hand to rowlock, find the force impelling the boat forward.

16. If a balance has unequal arms a and b and a shopman weighs alternately from each scale pan, does he ultimately lose or gain and how much?

17. If the span of a beam is 20 ft. and a load of 12 cwt. is shifted from one position through a distance of 5 feet along the beam, what difference in the reactions will this cause?

CHAPTER III

WORK, POWER AND ENERGY

THE term *work* is quite familiar to everybody but its general meaning is not very easy to express succinctly; it is used in mechanics in a special restricted sense and may for our purpose be defined as follows: *When a force acts upon a body and causes it to move it is said to do work on the body.* When the force is constant, work is measured by the product of the force and the distance through which the body moves in the direction of the force. The engineer's unit of work is the FOOT-POUND, i.e., the amount of work done by a constant force of one pound weight in moving a body through a distance of one foot in the direction of the force. If our forces are measured in tons and our distances in inches, our work will be in inch-tons and so on. If for instance the weight used in driving a clock weighs 20 lbs. and it drops through a vertical distance of 5 feet, the distance moved in the direction of the weight of the body, which acts vertically, is 5 feet. Therefore the work done by the weight is $20 \times 5 = 100$ ft.-lbs.

To express this idea generally, instead of by numerical illustration, suppose that a constant force F (Fig. 25) acts upon a body indicated by the shaded area situated originally at a point A and that after a certain time the body has been moved to a point B. Draw BC parallel to F and draw AC at right angles to it, then BC is the distance through which the body has moved in the direction of the force, so that the work done in the given time is measured by $F \times BC$.

It will be noticed that the work done by a force is measured by the product of a force into a length and that the moment of a force about a point is also measured by the product of a force into a length. In order to avoid confusion a distinction is

sometimes made in naming the compound units—thus work is
measured in foot-pounds, inch-tons, etc., and moments are
measured in pound-feet, ton-inches, and so on.

It is very important at this stage to note that if the force
is given in magnitude and direction the work done in the given
time depends only upon the original and final positions, A and
B respectively, of the body; it does not depend at all upon the
path taken.   The work done, for instance, for a straight line

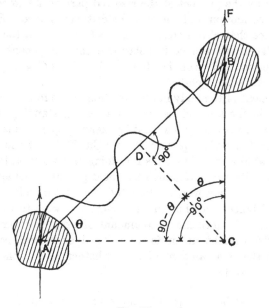

Fig. 25.

path between A and B is the same as for any curved path
such as that shown in the figure.

We will here note that the product of the force and the
distance moved in the direction of the force is exactly equal
to the product of the resolved part of the force in the direction
of motion and the actual straight distance moved.   To prove
this statement draw CD perpendicular to AB, and suppose
that a scale of force is so chosen that CB represents the force F.
Then DB represents the resolved part of F in the direction of
movement AB.

Then the work done $= F \times BC = BC^2$.

Now in the $\triangle$ ABC, $\sin \theta = \dfrac{BC}{AB}$, and in the $\triangle$ DCB, $\sin \theta = \dfrac{DB}{BC}$;

$$\therefore \frac{DB}{BC} = \frac{BC}{AB} \text{ or } DB \cdot AB = BC^2;$$

but DB is the resolved part of F in the direction of motion and AB is the distance moved so that the work done may also be measured by the product of the resolved part of the force in the direction of motion and the straight distance moved. Suppose for instance that F is 10 lbs. and AB is 6 inches and the angle θ is 30°. Then BC will be 3 inches and the component of F in the direction AB will be 5 lbs. So $F \times BC = 30$ inch-lbs. and $5 \times AB = 30$ inch-lbs.

Movement of the body in the direction of the force is essential for work in the scientific sense. If a man is standing still and is holding a heavy body, he must be exerting by muscular action a force on the body equal to the weight W lbs. of the body if it is also stationary; but he is not doing any work on the body in the scientific sense, although he would probably feel aggrieved if told so. If, however, he lifts the weight through a certain vertical distance x feet, his muscular effort does work to an amount Wx ft.-lbs. He does the same amount of work whether he lifts the weight straight up or in an inclined or curved path; he also does the same amount of work whether he lifts the weight quickly or slowly.

Power. If he lifts quickly, however, he exerts more power than if he lifts slowly, for *power is the rate of doing work*, i.e., power is the work done in a unit of time. The British unit of power is called the *Horse-Power* (H.P.) and is fixed at 33,000 *ft.-lbs. per minute* (or 550 ft.-lbs. per second). This unit was chosen by James Watt as the result of experiment with horses winding up weights. It is not a very satisfactory unit but it has become firmly established and it is now too late to alter it. [We have already used the idea of a 1000 lb. unit of force called the *kip*, so that a 1000 ft.-lbs. would be called a *foot-kip*. If we were setting out to choose a more convenient unit of power than the horse-power, we might take one foot-kip per second and call it the *Skip* which would be equivalent to

$\frac{1000}{550} = 1\cdot8181$ H.P. We do not propose, however, to adopt this unit throughout the book.].

Numerical Examples. (1) *What is the least power that a pump must be exerting when it is lifting water at the rate of* 1500 *gallons per minute through a vertical distance of* 50 *feet?*

A gallon of water weighs 10 lbs., so that the force exerted on the water is at least $10 \times 1500 = 15,000$ lbs. Actually the force will have to be a little more than this because of frictional and other resistances that have to be overcome; that is why the question is worded in the form given.

$$\text{The work done per minute} = 15,000 \times 50$$
$$= 750,000 \text{ ft.-lbs.}$$
$$\therefore \text{ Power} = \frac{750,000}{33,000}$$
$$= \underline{22\cdot7 \text{ H.P.}}$$

(2) *If the horse referred to in the example on p.* 7, *travels at the rate of* 3 *miles per hour for* 10 *minutes, how much work will he have done and at what horse-power will he be working?*

In ten minutes the horse will have walked half a mile, that is 2640 feet.

We have shown already that the component of the pull in the direction of motion of the horse is 141 lbs., so that the work done is $141 \times 2640 = 372,000$ ft.-lbs. One H.P. = 33,000 ft.-lbs. per min.=330,000 ft.-lbs. in 10 minutes.

$$\therefore \text{ Horse-Power} = \frac{372,000}{330,000} = \underline{1\cdot13}.$$

Energy. Energy is usually defined as the *capacity for doing work*. Energy exists in nature in several forms; thus we have electrical energy, light energy, heat energy, energy stored in water at high elevations, the principal source of all energy being the heat of the sun. Energy can be converted from one form into another and the principal function of the engineer is the conversion of natural energy into convenient forms for the benefit of man. Thus the energy stored up by the sun in bygone ages in the vegetation which has now become coal is converted in the boiler into the expansive energy of steam which drives steam-engines for performing countless kinds of mechanical work. In mechanics we divide mechanical energy into two kinds—*kinetic energy* and *potential energy*.

Kinetic energy (commonly written K.E.) *is the work which a body is capable of performing in virtue of its motion.* A familiar example of kinetic energy is that possessed by a bullet, which in being brought to rest can do a large amount of work; another example is that possessed by the wind, the kinetic energy of which has been employed from time immemorial to propel ships, and drive mills for the grinding of corn and for the pumping of water.

Potential energy is the work which a body is capable of performing in virtue of its position. A familiar example of this is given in an illustration which we have already considered, viz., a weight used in driving a clock. If the weight W lbs. is at a height h feet above the ground, it will do an amount of work equal to Wh ft.-lbs. before it comes to the ground; its potential energy is therefore said to be equal to Wh ft.-lbs. Another example is afforded by water at a high elevation which is often used to drive machinery to generate electric power.

In this connection we may point out that when in ordinary parlance we speak of power we really mean energy. An electric power station is really employed in generating electric energy, power meaning strictly, as we have already indicated, the rate of doing work.

The Conservation of Energy. We have already stated that energy can be *converted* from one form to another, but up to the present nobody has discovered a way of *creating or destroying* energy, nor do we think it probable that anybody ever will. Thus we get the doctrine which is the foundation of all physical science that " Energy can neither be created nor destroyed but can be converted from one form into another." This is known as the principle of the " conservation of energy." Failure to understand this law has led thousands of men to spend much valuable time and money in trying to invent perpetual motion machines, i.e., machines which when once started will go on working for ever without receiving any additional external energy. It is really quite remarkable that, although this doctrine has long been accepted by all scientists, there are still inventors who try to cheat nature of her laws and to make these machines.

As no energy is destroyed a given amount of heat energy can always be converted into the same amount of work. Thus one British Thermal Unit (B.TH.U.), which is the amount of heat required to raise the temperature of one pound of water through one degree Fahrenheit, has been found to be equivalent to 778 foot-pounds of work. This is commonly spoken of as the Mechanical Equivalent of Heat or Joule's Equivalent.

The commercial unit of electrical energy in this country is called the Board of Trade Unit (B.T.U.) and the unit of electrical power is the *watt* or the *kilowatt* (1000 watts); 1 kilowatt is one B.T.U. per hour (i.e., 1 kilowatt is the power which in one hour produces one B.T.U. of energy). Now 1 Horse-Power is equivalent to 746 watts; when therefore we know the amount of electrical energy used in a given time in any machine such as an electro-motor, we can tell exactly how many foot-pounds this is equivalent to.

Useful Energy. While it is true that energy cannot be destroyed, it is also true that in every conversion from one form of energy to another some of it is always wasted. Take for instance the case of a steam, gas or oil engine. A certain amount of energy is put into the engine in the form of steam or explosive mixture and a certain amount of work (called *useful* work) is done by the engine, but we can never use more than something like one-quarter of the amount of energy put in; of the remainder part escapes in the exhaust steam or gases and part is spent in overcoming the friction in the engine. The energy is not destroyed but much of it cannot be usefully employed; it is practically wasted. One pound of average coal contains about 12 million foot-pounds of energy; the greater proportion of this goes into the water in the boiler and the remainder goes up the chimney. It is a very good steam-engine that does not use more than 1½ lbs. of coal per H.P. hour.

Now   1 H.P. hour = 33,000 × 60 ft.-lbs.

$$= 1 \cdot 98 \text{ million ft.-lbs.}$$

1½ lbs. of coal contain 18 million ft.-lbs. of energy so that in a very good steam-engine 1·98 out of 18 or 11 per cent. of the energy supplied to it is usefully employed; the remaining 89 per cent. is wasted. The great problem to be faced by engineers of the future is that of obtaining mechanical energy in a less wasteful manner. An electro-motor wastes very much less energy than a steam-engine, but the electrical energy is nearly always obtained from coal by means of a steam-engine so that the energy of the coal is still to a large extent wasted. Gas and oil engines have now become less wasteful than steam-engines but even the best of them cannot give out as useful work more than one-third of the energy supplied to them

We shall return to this subject of wasted energy in the next chapter. In the meanwhile we will emphasize the fact that the conservation of energy is to be the basis of our treatment of mechanics. In any operation of a machine or action of a number of forces we will endeavour to find out what has become of the work that has been performed and by drawing up a kind of work balance-sheet we shall be able to investigate a number of points which are of the utmost importance in practice.

Work done by a variable force. In our examples illustrating the idea of work we have considered up to the present only the case in which the force is constant, but in most cases in

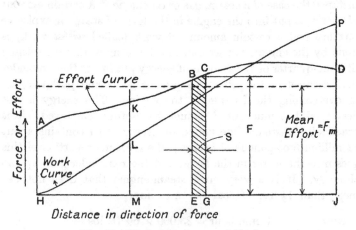

Fig. 26. Work done by a variable force.

practice the force varies from one time to another and if we based our ideas upon constant forces only we should not be able to deal clearly with the problems that arise in practice.

Suppose that the force acting upon a body in the direction

of its motion and driving it forward varies so that when we
plot a diagram of the force at various points we get a curve
ABCD, Fig. 26, which is usually called the *effort curve*. Consider
two points E and G which are so close together that the force
F may be regarded as constant over the length. Then the work
done over this length will be equal to constant force × distance
moved = F × S = area of the shaded strip of the curve. The
reader will see that the smaller we make the distance EG the
more nearly true will be the statement that the area of the strip
is equal to F × S, but that for comparatively long lengths the
statement is only approximate. If now we consider the whole
base HJ to be divided up into short lengths, the same argument
will hold for each strip of the curve so that adding together
these separate strips we see that the total work done in moving
the body from H to J is represented by the area HADJ.

Or we get the rule that:

*The work done is represented by the area beneath the effort
curve.*

Now suppose that we wish to find the work done up to various
points along the base and to obtain a diagram representing to
some other scale the work done. Such a diagram will be of the
form shown by the curve HLP in the figure and is called the
work curve. Consider any point L on this curve. Then the
ordinate LM represents the work done in moving from H to M
and this is also given, as we have proved above, by the area
HAKM. We see therefore that the ordinate of the work curve
at any point represents the area of the effort curve up to the
same point. When two curves have this relation, the first is
said to be the *sum curve* of the second; thus in our case the
work curve is the sum curve of the effort curve. A graphical construction for the sum curve is given in the appendix (p. 294).

Numerical Example. *The force urging a body forward
increases uniformly from zero to 2000 lbs. during the first 15 feet
of movement; it then remains constant for the next 20 feet; and
finally decreases uniformly to zero in a further 20 feet. Find
the work done and the constant force which would do the same
amount of work in moving the body through the same distance.*

Fig. 27 shows the effort curve in this case. Therefore the
work done is given by the area of the figure ABCD.

This is equal to

$$\text{ft.-lbs.}$$

Area of $\triangle ABE$ $= \frac{1}{2} \times 15 \times 2000 = 15,000$
Area of rectangle $BCFE = 20 \times 2000 \quad\quad = 40,000$
Area of $\triangle CDF$ $= \frac{1}{2} \times 20 \times 2000 = 20,000$

Total work done $= \underline{75,000}$ ft.-lbs.

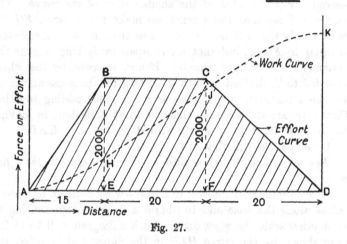

Fig. 27.

The total distance moved is 55 feet. If therefore a constant force F were acting we should have $55F = 75,000$;

$$\therefore F = \frac{75,000}{55} = \underline{1364 \text{ lbs. nearly.}}$$

We show in dotted lines in Fig. 27 the work curve for this case. The portion AH is a parabola with vertex at A; HJ is a straight line, and JK is a parabola with vertex at K. The student should make this construction as an exercise. Take for instance as scales: Distance $1'' = 10$ feet. Force $1'' = 1000$ lbs. Polar distance $p = 2 \cdot 5$ actual inches.

Then the work scale will be $1'' = 2 \cdot 5 \times 10 \times 1000 = 25,000$ ft.-lbs.

DK should therefore be 3 inches.

Work against Resistance. In every case that arises in practice there is a force resisting the movement of a body under the driving force or effort, such resisting force is called the Resistance or sometimes the "external resistance."

Take the case of a steamboat, Fig. 28. The steam acting
upon the pistons and thence upon the propeller causes a certain
tractive effort F to be exerted tending to push the steamer
forward; the resistance of the water and other external forces
tending to resist the forward motion of the steamer cause
a resistance force S to be exerted in the opposite direction.
If F is greater than S at any instant, work will be done upon the
steamer and as such work cannot be lost it becomes converted
into increased kinetic energy, and if F is less than S the kinetic
energy of the steamer will decrease; this is expressed in simple
language by saying that, if F is greater than S the speed of the
vessel will increase, but if S is greater than F the speed will

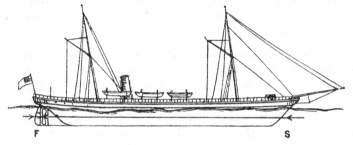

Fig. 28.

decrease. The kinetic energy can be regarded as energy stored
up for use in emergency; if the effort is less than the resistance,
the body gives up some of its kinetic energy to make up the
difference between the work done by the effort and the resistance.
When this difference in work is equal to the whole kinetic energy
that the body possessed in the first place, the body will stop
moving.

This is the first time that we have dealt with the case in
which a body may move in a direction opposite to that in which
the resulting force upon it acts. When the direction of movement is opposite to that of the force we shall speak of the force
as taking work from the body.

Resistances are nearly always what may be called "induced"
or "passive"; that is to say they disappear directly the body
comes to rest. The resistance to the motion of a steamer increases
very quickly with the speed and we soon get to a speed which
may be regarded as the most economical. A slight increase of

speed over this will require more coal and cost more money than the saving in time is worth.

We have another similar example in racing motor cars. An ordinary 12 H.P. car can do 30 miles per hour but to get 80 miles an hour we have to increase the horse-power to something like 80 or more.

In nearly everything that engineers have to deal with, it is energy that they must try to use to the best advantage, because money is merely the token for energy. If the world's supply of coal, oil and other fuel gave out, it would take very few years before we should nearly all be starved to death. James Watt's improvements of the steam-engine probably did more for the benefit of humanity than any scheme that human skill has devised, because it opened up vast fields for the use of the energy stored up in fuel.

Graphical representation of Effort and Resistance. Suppose that the effort in moving a body from a point X to a point T varies in the manner indicated by the curve ABC, Fig. 29, and that the resistance varies in the manner indicated by the curve DBF. Then if we take two points KL very close together on the base—so close that the effort F and resistance S may for all practical purposes be considered as constant over the length— the work done upon the body by the effort from K to L is equal to force × distance $= F \times KL =$ area of strip $EGLK$. Therefore as already shown the total work done on the body by the effort in moving from X to T is equal to the area $ABCTX$.

Similarly the work taken by the resistance from the body in moving from K to L is equal to $S \times KL =$ area of strip $HJLK$; so that the total amount of work taken from the body in moving from X to T is equal to the area $DBFTX$.

Now the resultant work on the body is equal to work done by the effort − work taken away by the resistance

$$= \text{area } ABCTX - \text{area } DBFTX$$
$$= \text{area } ABD - \text{area } BFC.$$

At any intermediate point such as K the excess of work done by the effort over the work expended in overcoming the resistance is the difference between the areas $XAEK$ and $XDHK$.

Therefore between the points X and U the body increases in kinetic energy by the amount represented by the area ADB and it then loses in going between U and T an amount of kinetic energy represented by the area BFC. We have not yet explained how the kinetic energy can be expressed in terms of the velocity but we have seen that the velocity is an indication of the kinetic energy; consequently at the point U the body has the maximum amount of kinetic energy and therefore has the maximum velocity or speed.

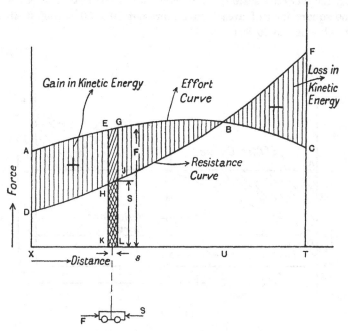

Fig. 29. Work against Resistance.

If the conditions were reversed so that the resistance were at first greater than the effort and less at the end, the body would be losing kinetic energy up to the point U; U would then be the point of least kinetic energy and therefore of least velocity.

We shall deal later with many problems concerned with kinetic energy; our present aim is just to make clear the idea of work and energy and the fact that energy is never destroyed.

Numerical Example of Effort and Resistance. *A body is being urged forward by a constant force equal to 100 lbs. and over a distance of 120 feet the resistance increases uniformly from 30 lbs.*

to 150 *lbs. At what point will the body move with the greatest velocity? How much kinetic energy will the body then have gained and how much will it have gained at the end of the* 120 *feet ?*

Referring to Fig. 30 *ABC* is the effort curve and *DBF* is the resistance curve.

The point *U* gives the point of maximum velocity. The distance *XU* can be measured by drawing the diagram to scale, e.g. distances to a scale $1'' = 20$ feet and forces to a scale $1'' = 50$ lbs.; one square inch of area would represent $20 \times 50 = 1000$ ft.-lbs. It will come to 70 feet.

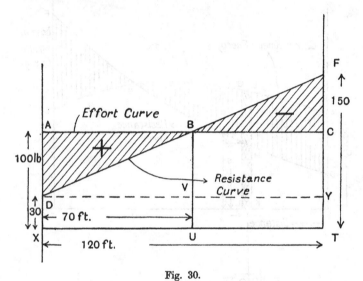

Fig. 30.

By calculation we should proceed as follows. Draw *DY* horizontally as indicated in dotted lines.

Then $\dfrac{DV}{DY} = \dfrac{BV}{FY}$ because *BV* is parallel to *FY*.

$$\therefore DV = \frac{DY \cdot BV}{FY} = \frac{120\,(100 - 30)}{150 - 30} = \frac{120 \cdot 70}{120} = \underline{70 \text{ feet.}}$$

Gain in K.E. up to U = area of $\triangle ABD = \tfrac{1}{2}AB \cdot AD$
$$= \tfrac{1}{2} \cdot 70 \cdot 70$$
$$= \underline{2450 \text{ ft.-lbs.}}$$

Gain in K.E. up to T = area of $\triangle ABD$ – area of $\triangle BFC$

$$= 2450 - \tfrac{1}{2}.50.50$$
$$= 2450 - 1250$$
$$= \underline{1200 \text{ ft.-lbs.}}$$

Mean Effort. It is sometimes convenient to find the uniform effort which acting over the same distance will do the same amount of work as a variable one; this is called the *mean effort*. Referring to Fig. 26 let F_m be the mean effort; then work done by $F_m = F_m \times HJ$. But the work done by the mean effort has to be equal to the work done by the variable effort.

$$\therefore F_m \times HJ = \text{area } HAKDJ,$$

$$\therefore F_m = \frac{\text{area } HAKDJ}{HJ}.$$

Expressing this in general terms we have

$$\text{Mean effort} = \frac{\text{Area below effort curve}}{\text{Length of effort curve}}.$$

SUMMARY OF CHAPTER III.

The **work done** by a force upon a body is measured by the product of the force and the distance through which the body moves in the direction of the force. The unit is the **foot-pound**.

The work depends only on the initial and final positions of the body and not upon the path taken between the points.

Power is the rate of doing work, i.e., the number of foot-pounds of work done per unit of time.

One Horse-Power is equivalent to 33,000 foot-pounds per minute.

Energy is the capacity for doing work and can exist in various forms which can be converted from one to the other.

Mechanical energy can be divided into two kinds: *kinetic energy* (energy of motion) and *potential energy* (energy of position).

The law of the **conservation of energy** states that while energy can be converted from one form into another, it can be neither created nor destroyed.

If the force or effort be plotted against the distance, the result is called the *effort curve* and the work done up to any point is represented by the area of the effort curve up to that point.

The work curve is the sum curve of the effort curve.

In doing work upon a body against a resistance, the difference between the work done by the effort and the work done against the resistance goes in changing the kinetic energy of the body.

EXERCISES. III.

1. A chain 200 yards long and weighing 6 lbs. per ft. hangs vertically down a mine shaft. Find the work done in hauling it to the surface.

2. If in the preceding a weight of $\frac{1}{2}$ ton is attached to the end of the chain, find the total work done. Express each of the above results by a diagram.

3. Find the horse-power of an engine which will lift the weight in Question 2, in 25 seconds.

4. Find the horse-power required to pump 3000 gallons of water from a depth of 250 ft. in 10 minutes.

5. How many cubic ft. of water would an engine working at 100 H.P. raise per min. from a depth of 25 fathoms?

6. Find the work done in excavating a circular well 8 feet diameter, 45 feet deep, the weight of 1 cubic yard of earth being 1 ton. Give answer in ft.-lbs.

7. A horse drawing a cart at the rate of 2 miles per hour exerts a tractive force of 156 lbs. weight. Find the work done in 1 minute.

8. How many horse-power would be required to raise 2000 cubic feet of water per hour from a mine whose depth is 180 fathoms?

9. Find the horse-power required to draw a train along a level at 45 miles per hour, whose weight is 250 tons, the resistances being taken at 15 lbs. wt. per ton.

10. A cage with coals together weighing 10 cwt. is carried on the end of a wire rope weighing 10 lbs. per yard. Find the work done in ft.-lbs. in lifting it from the bottom of a mine 1500 ft. deep.

11. The travel of the table of a planing machine which cuts both ways is 9 ft. If the resistance to be overcome while cutting be taken at 400 lbs. and the number of double strokes per hour be 80, find the H.P. absorbed in cutting.

12. When a prismatic column of stone, 20 ft. diameter outside, 10 ft. diameter inside, 90 ft. high is being built, what actual work is done in lifting the stone from the ground? One cubic ft. of stone weighs 125 lbs.

13. What must be the effective H.P. of a locomotive which moves at the steady speed of 35 miles an hour on level rails the weight of the engine and train being 120 tons and the resistances 16 lbs. per ton? What additional H.P. would be necessary if the rails were laid along a gradient of 1 in 112?

14. Each of the two cylinders in a locomotive engine is 16″ diameter and the length of crank is 1 ft. If the driving wheels make 105 revolutions per minute and the mean effective steam pressure is 85 lbs. per sq. in., what is the H.P.?

15. A chain hanging vertically 520 ft. long weighing 20 lbs. per ft. is wound up. What work is done?

16. A 10-ton hammer falls through a height of 6 feet and makes an impression on a mass of iron to the extent of 1 in. Find the mean statical pressure in tons which has been exerted on the mass of iron during the blow.

17. A body weighing 1610 lbs. was lifted vertically by a rope, there being a damped spring balance to indicate the pulling force F lb. of the rope. When the body had been lifted x ft. from its position of rest, the pulling force was automatically recorded as follows:

x	0	11	20	34	45	55	66	76
F	4010	3915	3763	3532	3366	3208	3100	3007

Find approximately the work done on the body when it has risen 70 ft. How much of this is stored as potential energy and how much as kinetic energy?

CHAPTER IV

MACHINES AND EFFICIENCY

A machine may be described as an appliance for receiving energy from some outside source and delivering it in some more convenient form for doing work. Almost the simplest possible form of machine is the lever, which in the form of the crow-bar is used for lifting heavy packing cases. A man unaided cannot move the case; that is to say he cannot exert a force sufficiently great to lift it. He possesses quite enough energy to do so, but he can exert only a comparatively small force to move a body, although he can continue to exert it over a long distance; whereas for lifting the case he requires to exert a large force over a short distance; the crow-bar enables him to do this.

We have another every day illustration in the use of two and three speed gears in bicycles. When the cyclist comes to a hill, he puts in the low gear. This does not give him any more energy, in fact it makes him lose a little more than usual on account of the extra complication of the mechanism, but it enables him to use his energy more conveniently. He goes more slowly up the hill but does not have to push so hard and he finds that the result is a gain in comfort.

Wheel and Axle. We have a very simple form of machine in the wheel and axle shown in Fig. 30 a and we will show that we can get the same result by considering the work done as by considering the moments of the forces acting. In many problems a consideration of work done gives the quickest results.

A weight W to be lifted is connected by a rope or chain to an axle B of radius r which is supported in bearings and carries a wheel C of radius R to which the effort F is applied.

By moments we should have $Wr = F \cdot R$,

i.e. $$F = \frac{Wr}{R}.$$

Now suppose that the axle makes one revolution.

The weight W moves up a distance $2\pi r$ and the work done upon the weight is therefore $W.2\pi r$.

The rope to which F is applied moves downwards a distance $2\pi R$ so that the effort does an amount of work equal to $F.2\pi R$. If the axle runs freely, these two amounts of work must be equal.

$$\therefore F.2\pi R = W.2\pi r,$$

or $$F = \frac{Wr}{R},$$

as before.

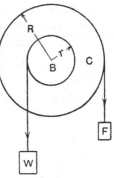

Fig. 30 a.

Crow-bar. Referring to Fig. 31, the full lines indicate the position of the crow-bar after the case has been raised while the dotted lines indicate the position before raising. F is the effort that the man can exert upon the end A of the lever and S is the resistance which the lever exerts upon the case at the point B. If C is the pivot or fulcrum of the lever and the perpendicular distances from C to the lines of action of F and S are respectively x and y, we have seen already that by the principle of moments

$$F \times x = S \times y \ \dots\dots\dots\dots\dots(1).$$

If therefore $F = 100$ lbs.; $x = 30$ inches and $y = 2$ inches,

$$S = \frac{Fx}{y} = \frac{100 \times 30}{2} = \underline{1500 \text{ lbs.}}$$

Mechanical Advantage. In a machine for converting one form of mechanical energy into another the ratio $\dfrac{\text{Resistance}}{\text{Effort}}$ is called the *mechanical advantage*.

In our particular case above we have

$$\text{Mechanical advantage} = \frac{S}{F} = \frac{1500}{100} = 15.$$

It will be noted that the "arm" x of the effort F in the position shown in full lines is appreciably larger than in the original position shown in dotted lines and that the arm "y" of the resistance S does not change appreciably; this shows that the mechanical advantage of this particular machine increases as the weight is lifted, so that the effort F will gradually diminish.

Now let us suppose that the distance A_1A_2 is so small that F and S are for all practical purposes constant during the lifting action.   Then we have

$$\text{Work done by effort} = F \times h.$$
$$\text{Work done by resistance} = S \times z.$$

If no work is wasted by frictional forces at the fulcrum or pivot, these two amounts of work must be equal or

$$Fh = Sz \dots\dots\dots\dots\dots\dots (2).$$

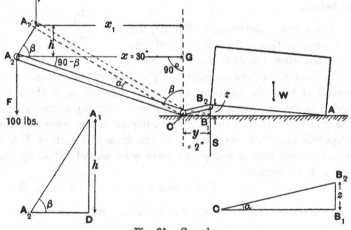

Fig. 31.   Crow-bar.

We have already shown by (1) that

$$Fx = Sy.$$

Therefore dividing we get

$$\frac{h}{x} = \frac{z}{y} \dots\dots\dots\dots\dots\dots\dots(3).$$

We will now try and prove that this should be the case; the student should try to verify this by drawing to scale and measuring to length. This is particularly desirable because the proof is rather long. We have already explained that the distance A_1A_2 is very small; therefore the angle a will be small; so small that the line A_1A_2 will not be appreciably different in length from an arc with centre C.

Now angle = arc ÷ radius;

$$\therefore a = \frac{A_1 A_2}{A_2 C} \text{ and also } a = \frac{B_1 B_2}{CB_1} = \frac{z}{y};$$

$$\therefore \frac{z}{y} = \frac{A_1 A_2}{A_2 C} \quad\dotsb(4).$$

Again the angle $A_1 A_2 C$ is practically a right angle; $\therefore GCA$ which is $90° - GA_2 C$ is equal to β; and

$$\therefore \sin \beta = \frac{h}{A_1 A_2} \text{ and also } = \frac{GA_2}{CA_2}, \text{ that is } = \frac{x}{CA_2};$$

$$\therefore \frac{h}{A_1 A_2} = \frac{}{CA_2} \text{ or } \frac{A_1 A_2}{A_2 C} = \frac{h}{x} = \frac{z}{y} \text{ (from (4))}.$$

This is the result that we attempted to prove. We see therefore that the principle of moments gives us the same result as the principle of work; in some problems it is more convenient to use moments and in others it is more convenient to consider the idea of work.

Efficiency of Machines. It is of considerable help in the understanding of the mechanical principles of machines to imagine a machine to be a kind of box, as indicated in Fig. 32,

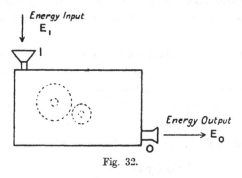

Fig. 32.

provided with an inlet I into which a certain amount of energy E_I is put in in a given time and with an outlet O through which issues an amount E_O of energy in the same time. Now in every machine a certain amount E_W of energy is lost or wasted. Therefore we may write Inlet Energy = Outlet Energy + Waste Energy.

Or in symbols $E_I = E_O + E_W$ \quad\dotsb(5).

Now the quantity $\dfrac{E_O}{E_I}$ is called the *Efficiency* of the machine.

Expressing this in words we should say that *the efficiency of a machine is the ratio of the energy that it gives out to the energy that it receives.*

Suppose, for instance, that a certain machine receives 120,000 ft.-lbs. of energy in a certain time and that during the same time it gives out 97,000 ft.-lbs.

Then   Efficiency $= e = \dfrac{97,000}{120,000} = \underline{\cdot 808 \text{ nearly.}}$

It is the usual practice to express efficiencies as so much per cent., i.e., to multiply the actual efficiency by 100. In our case, therefore, we should then say that the efficiency is 80·8 per cent.

Again since the energy wasted $E_W = E_I - E_O$ we have

$$\frac{E_W}{E_I} = 1 - \frac{E_O}{E_I} = (1 - e) ;$$

$\therefore$ energy wasted $= (1 - e) \times$ energy input.

The highest possible efficiency that a machine can have is 1 or 100 per cent. but most machines have an efficiency considerably less than this, the simpler machines generally having a higher efficiency than the more complicated ones. The principal aim that an engineer has in designing machines is to make the efficiency as high as possible, that is to make the energy wasted as small as possible.

Velocity ratio of Machines. In a machine in which the effort and resistance are constant in direction, the quantity

$$\frac{\text{Distance moved at the effort in a given time}}{\text{Distance moved at the resistance in the same time}}$$

is called the *velocity ratio.*

In the example shown in Fig. 31 we should have

$$\text{Velocity ratio} = \frac{h}{z} = V_r .$$

In this case we showed that the mechanical advantage if there was no loss was

$$\frac{S}{F} = \frac{x}{y} = \frac{h}{z} .$$

Therefore when there is no loss of energy mechanical advantage = velocity ratio.

IV] MACHINES AND EFFICIENCY 57

When, however, some energy is lost we have to modify this result because, although the velocity ratio is fixed by the actual sizes of the elements forming the machine, the mechanical advantage depends upon the amount of energy that is wasted. As a more general rule therefore we say that

Mechanical advantage = velocity ratio × efficiency,

and since $\qquad$ Mechanical advantage $= \dfrac{\text{Resistance}}{\text{Effort}}$,

we may say that

Resistance = effort × velocity ratio × efficiency,

i.e. $\qquad\qquad S = F \cdot V_r \cdot e.$

In the case of the crow-bar that we have examined the velocity ratio is 15. Now suppose that the pivot is rough so that energy is absorbed in moving it and suppose that 3 % of the energy input is wasted, then efficiency $= e = 1 - \frac{3}{100} = \cdot 97$.

Then we should have

$$S = 100 \times 15 \times \cdot97 = 1455 \text{ lbs.,}$$

or if we require to find the value of F for $S = 1500$ we have

$$1500 = F \times 15 \times \cdot97;$$
$$\therefore F = 103 \text{ lbs.}$$

Now the effort that would be required in a perfect machine, in which no energy is wasted and whose efficiency is therefore 1, we shall call the *ideal effort*.

Now $\qquad$ Actual effort $= \dfrac{\text{Resistance}}{\text{Mechanical advantage}}$,

and $\qquad$ Ideal effort $= \dfrac{\text{Resistance}}{\text{Velocity ratio}}$;

$$\therefore \frac{\text{Ideal effort}}{\text{Actual effort}} = \frac{\text{Resistance}}{\text{Velocity ratio}} \div \frac{\text{Resistance}}{\text{Mechanical advantage}}$$

$$= \frac{\text{Mechanical advantage}}{\text{Velocity ratio}} = \text{Efficiency,}$$

i.e. $\qquad$ Actual effort $= \dfrac{\text{Ideal effort}}{\text{Efficiency}}$.

Some simple Machines. The inclined plane. The inclined plane is one of the most ancient forms of machine and is one of the simplest. Suppose, for instance, that we wish to raise a body such as a truck up to a point B. It is too heavy to lift, but by running an inclined plane AB from the ground level to the point we can push the body slowly up.

(a) *Effort parallel to the plane.*

If F (Fig. 33) is the effort or force pushing the body up the plane, and acting parallel to it, the work done by the effort in moving the body up is $F \times AB$. The body has to be raised a vertical distance BC; this is the distance moved in the direction of its weight, representing an amount of work equal to $W \times BC$.

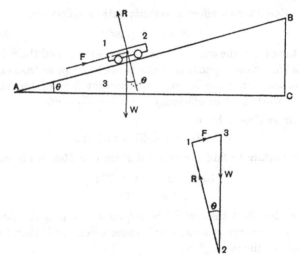

Fig. 33. Inclined Plane. Effort parallel to Plane.

If the velocity is constant between A and B there is no change in kinetic energy and if also there is no energy wasted we have

Work done by effort F = Work done against weight W;

$$\therefore F \times AB = W \times BC,$$

or $\quad\quad \dfrac{W}{F} = \dfrac{AB}{BC}$ = Mechanical advantage.

Also   Velocity ratio = $\dfrac{\text{Distance moved in direction of } F}{\text{Distance moved in direction of } W}$

$$= \frac{AB}{BC}.$$

We can also find a relation between F and W by considering the forces acting upon the body. The third force is the reaction

R between the truck and the plane; if there is no friction this reaction will be at right angles to the plane. Therefore by drawing lines parallel to the forces to a convenient scale we have a triangle of forces or vector figure 1, 2, 3.

It will be noted that each side of this triangle is at right angles to a side of the triangle ABC; therefore the two triangles are similar, therefore

$$\frac{W}{F} = \frac{2, 3}{2, 1} = \frac{AB}{BC},$$

or

$$F = \frac{W \cdot BC}{AB} \quad \ldots\ldots\ldots\ldots\ldots (1).$$

Also

$$\frac{R}{W} = \frac{1, 3}{2, 3} = \frac{AC}{AB};$$

$$\therefore R = W \cdot \frac{AC}{AB} \quad \ldots\ldots\ldots\ldots (2).$$

We can use the language of trigonometry to express these results as follows:

$$\frac{BC}{AB} = \sin \theta;$$

$$\therefore E = W \sin \theta \ldots\ldots\ldots\ldots\ldots (3),$$

$$\frac{AC}{AB} = \cos \theta; \quad \therefore R = W \cos \theta \quad \ldots\ldots\ldots (4).$$

We also note that

$$\tan \theta = \frac{BC}{AC}.$$

Numerical Example. *What force is necessary to push a truck weighing 15 tons up a gradient of 1 in 10 ?*

This means that BC is 1 when AC is 10.

Then since $\qquad AB^2 = BC^2 + AC^2,$

$$AB = \sqrt{1 + 100} = \sqrt{101} = 10\text{·}05 \text{ nearly};$$

$$\therefore F = \frac{15 \times 1}{10\text{·}05} = \underline{1\text{·}49 \text{ tons.}}$$

In practice F will always be more than this because there is always some energy wasted, the principal cause of waste being called *friction*. The above value of F is the *Ideal effort*.

Experiment. A very simple piece of apparatus can be rigged up for the experimental verification of the laws of the inclined plane. It is constructed by fixing a board C (Fig. 34) by a hinge D to one end of a board A. A vertical board E is fixed at the end B and is provided with a slot through which passes a shouldered pin provided with a fly-nut G so arranged as indicated in the detailed figure that the board C rests on the projecting pin. A scale S is carried by the vertical board E and a pulley P is fixed in the free end of the board C.

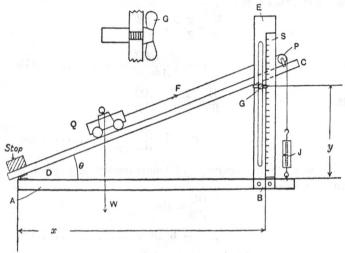

Fig. 34.

A light string is attached to a truck Q in which is placed a weight, the combined weight of the truck and weight being equal to W; the string passes over the pulley P and through a slot in the end of the board C and has its other end attached to a spring balance J.

We then measure carefully the distance x from the hinge to the edge of the scale and write it at some convenient place on the apparatus.

The pin is then set to a certain value of y, the fly-nut tightened up, and the reading on the spring balance is noted.

If we wish to save time in our calculations by using trigonometrical methods we next proceed to calculate the value of the angle θ for various values of y and draw a calibration diagram.

Suppose for instance that x = 30 inches.

When y = 5 inches we have tan θ = $\frac{5}{30}$ = ·1667; from trigonometrical tables we find that θ = 9·5 degrees.

When y = 10 inches we have tan θ = $\frac{10}{30}$ = ·3333, and we get by tables θ = 18·5 degrees about.

Similarly we get   y = 15   θ = 26·6 degrees.
                   20   = 33·7   „
                   25   = 39·8   „
                   30   = 45·0   „

We then by plotting obtain the calibration diagram shown in Fig. 35.

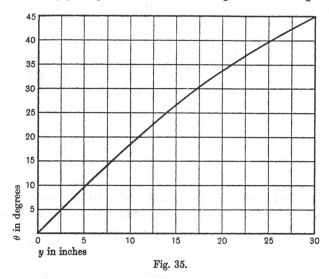

Fig. 35.

From this diagram we can read off at once the angle θ for any value of y and can calculate the theoretical values of F. The results can then be tabulated as follows:

W lbs.	y inches	Spring balance reading = F lbs.	θ [from diagram]	Ideal value of F = $W \sin \theta$ lbs.

If we wish to avoid trigonometry we note that the quantity corresponding to AB in Fig. 33 = $\sqrt{x^2 + y^2}$;

$$\therefore \text{ Theoretical value of } F = \frac{W \cdot y}{\sqrt{x^2 + y^2}}.$$

The above gives what is called a *static* test. To get an approximation to a *dynamic* or *running* test, take away the spring balance and replace it by a scale pan. Let the truck rest against the stop and weight the pan until the truck moves up slowly without gaining in speed; then the combined weight of scale pan and weights gives the effort F required to move the truck up without increasing its speed and therefore its kinetic energy. Now take the weights off slowly until the truck begins to run back without increasing its speed; the resulting weight is the effort that the truck can exert in moving down. We can then tabulate conveniently as follows:

W lbs.	y inches	θ [from diagram]	Ideal value of F $= W \sin \theta$ lbs.	Observed value of F in lbs.	
				Rising	Falling

(b) *Effort parallel to the ground.* Now let the force F act horizontally as indicated in Fig. 36. When the body has moved

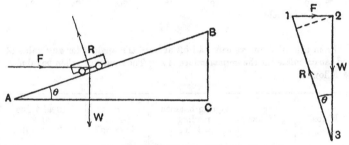

Fig. 36. Inclined Plane. Effort horizontal.

from A to B, it has gone in the direction of the effort F a distance AC and in the direction of the weight W a distance CB. If therefore no energy is wasted

Work done by effort F = Work done against weight W

or $$F \times AC = W \times BC,$$

i.e. $$\frac{W}{F} = \frac{AC}{BC} = \text{Mechanical advantage};$$

$$\therefore F = W \cdot \frac{BC}{AC} = W \tan \theta.$$

Considering the subject from the standpoint of equilibrium of forces we have 1, 2, 3 as the vector figure or triangle of forces. It will be noted that this triangle is the same as the triangle

ABC drawn to a smaller scale and turned through ninety degrees.

Therefore the triangles are similar or

$$\frac{F}{W} = \frac{1,\,2}{2,\,3} = \frac{BC}{AC};$$

$$\therefore F = \frac{W \cdot BC}{AC}, \text{ as before.}$$

Also

$$\frac{R}{W} = \frac{1,\,3}{2,\,3} = \frac{AB}{AC}.$$

The dotted line on the vector figure shows the value of F in case (a).

The Screw. The screw is a form of inclined plane, and may be considered as an inclined plane wrapped round a cylinder.

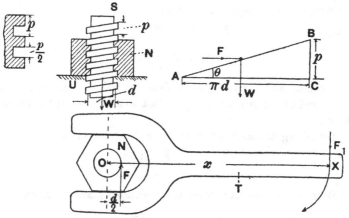

Fig. 37. The Screw.

It is formed by cutting a groove in the cylinder leaving a projection or *thread* which may be either of triangular or square form. In Fig. 37 is shown a square thread. It is engaged by a *nut N* having a corresponding thread in it. If the nut is fixed and the screw shown is turned in a clockwise direction the screw moves further down into the nut; such a screw is called *right-handed*. If on rotating the screw in a clockwise direction it moved upwards out of the nut, it would be called *left-handed*. The distance moved into or out of the nut in one turn is called the *pitch* of the screw; the pitch might also be defined as the distance parallel to the axis of the screw between corresponding edges of two successive threads.

If we consider one thread of the screw as unwound until it is all in one plane we should get the inclined plane ABC.

Now suppose that the screw supports a weight W and that it is prevented from rotating; if the nut is turned in a clockwise direction by means of a spanner T, a force F_1 being applied at a distance x from the centre of the screw, the weight will be lifted.

Then if F_1 is altered in direction so that it is always at right angles to the spanner as the latter rotates, the distance moved in the direction of F_1 in one revolution will be equal to the circumference of a circle of radius x, i.e. $2\pi x$, so that the work done by F_1 is equal to $F_1 \times 2\pi x$. In this one revolution of the nut the weight will be lifted by an amount p so that the work done on the weight will be equal to $W \times p$;

$\therefore$ we have $\qquad F_1 \times 2\pi x = W \times p$

or $\qquad\qquad F_1 = W \cdot \dfrac{p}{2\pi x}.$

We may also consider the problem as follows; the force F_1 at the end of the spanner is equivalent to a force F at the screw thread.

By taking moments about O, the axis of the screw, we have

$$F_1 \times x = \frac{F \times d}{2}.$$

By our previous treatment of the inclined plane we have

$$F_1 = W \tan \theta = \frac{Wp}{\pi d};$$

$$\therefore F_1 \times x = \frac{Fd}{2} = \frac{Wp}{\pi d} \cdot \frac{d}{2}$$

$$= \frac{Wp}{2\pi};$$

$$\therefore F_1 = \frac{Wp}{2\pi x}, \text{ as before.}$$

This is of course the ideal effort; in practice it will be more on account of the energy wasted due to the friction between the nut and the thread and between the nut and the fixed surface U.

Screw-Jack. A very common form of machine employing the screw is the "screw-jack" which is used for lifting heavy bodies through short distances and is used largely for lifting motor cars at one side in order to remove the wheel.

It consists of a screw A (Fig. 38) working in a nut formed in the top of a base B, the end of the screw ending in a knob portion provided with "tommy-holes" D, the extreme end being turned to a smaller diameter and carrying a thrust cap E provided with

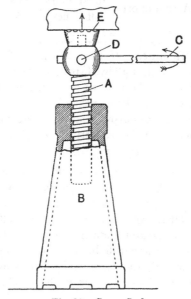

Fig. 38.   Screw-Jack.

ridges to give a good grip. A hand-lever C is passed through one of the tommy-holes and is pushed round in the direction of the arrow; when the lever has been given about a quarter turn, it is put through the next hole and pushed round further thus slowly raising the article to be lifted, the nut being fixed.

Numerical Example. *A screw-jack has the screw of $\frac{1}{2}''$ pitch and the lever is 15 inches in length from the centre of the screw to the point at which it is grasped. What force must be exerted on the lever to lift a load of 2 tons if the efficiency of the machine is 40 %?*

Suppose that the lever makes one complete turn.

Distance moved by effort $= 2\pi \times 15$ inches.

Distance moved by resistance or weight = pitch of screw = $\frac{1}{2}$ inch.

$$\therefore \text{Velocity ratio} = \frac{2\pi \times 15}{\frac{1}{2}} = 60\pi = 188\cdot5\,;$$

$$\therefore \text{Ideal effort} = \frac{\text{Resistance}}{\text{Velocity ratio}}$$

$$= \frac{2 \times 2240}{188\cdot5} = 23\cdot8\,\text{lbs.}$$

$$\therefore \text{Actual effort} = \frac{\text{Ideal effort}}{\text{Efficiency}} = \frac{23\cdot8}{\cdot4}$$

$$= 59\cdot5\ \text{lbs.}$$

Reversing Machines. If the efficiency of a machine be sufficiently great it will, if allowed, reverse, that is the resistance acting as an effort will be able to make the machine run backward, but if the efficiency be less than 50 % this cannot happen.

For let the input, the output, and the waste energy, when the machine is acting direct, be respectively E_I, E_O, and E_W, then

$$E_I = E_O + E_W.$$

Now let the resistance act as an effort and do work E_O, the body moving through the same distance as before; the amount of waste energy is again E_W and the balance $E_O - E_W$ will be available as output at what was originally the effort end of the machine. If E_O is greater than E_W, that is if the efficiency is greater than 50 %, there will be some work delivered, but if E_W is greater than E_O the resistance will not even be able to overcome the wasteful forces, that is the machine cannot run back unaided.

This general explanation may be a little difficult to follow at first but will probably be made clear by the following numerical illustration.

Suppose that we have a machine with velocity ratio 10 and efficiency $\cdot4$ and let the resistance be 100 lbs.

Also let the part of the machine at which the effort is applied move through 10 feet in the direction of the effort; then the resistance end moves through 1 foot in the direction of the resistance.

Then   Input energy E = 25 × 10 = 250 ft.-lbs.,

Output energy E_O = 100 × 1 = 100 ft.-lbs.;

$\therefore$ Waste energy E_W = 250 - 100 = 150 ft.-lbs.

Now let us reverse the conditions and allow 100 lbs. to act as an effort through 1 foot if it can; it would do 100 ft.-lbs. of work which is not sufficient to supply the 150 ft.-lbs. of waste energy so that 100 lbs. will not be sufficient to reverse the machine.

A machine that will not reverse is called *self-sustaining*. In some machines this is a convenience; for instance in the case of the screw-jack previously described. In such cases we have to pay for the convenience by low efficiency.

Pulley Tackle. The various forms of pulley tackle are examples of simple forms of machine.

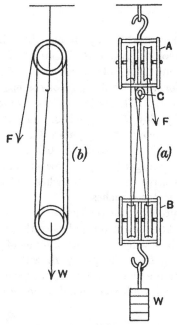

Fig. 39. Pulley Tackle.

Fig. 39 (*a*) shows one form. It consists of two blocks *A*, *B* each consisting of two pulleys of equal size. The rope is fixed to an eye *C* in the upper block and then passes over one pulley in the lower block; then over one of the pulleys in the upper block; then over the other pulley in the lower block and finally over the remaining pulley in the upper block. Fig. 39 (*b*) shows the arrangement diagrammatically, the two pulleys in each

5—2

block being of slightly different diameters to show more clearly
the manner in which the rope passes over them.

Now suppose that the rope at the end F is moved downwards
one inch, the lower block will then move upwards $\frac{1}{4}$ inch because
there are four ropes that have to move up by the same amount
and the total amount of upward movement must be equal to
the downward movement at the end F because the rope is
continuous.

If therefore there is no loss of energy we shall have

$$F = \frac{W}{4}.$$

Weston's Differential Pulley Block. This block consists of
two specially grooved pulleys, A, B, Fig. 40 (a), of slightly different
diameters cast in one piece and secured to a strong upper support.
The grooves are formed with flat portions to engage the chain in
the manner of teeth. The weight W is secured to a second
similarly-grooved pulley C. An endless chain F passes over the
larger pulley A; then over the pulley C; and then over the
smaller pulley B, as shown, the effort F being applied to the chain
that comes over the larger pulley. Now suppose that the larger
pulley A is of effective diameter D inches and that the smaller
is of effective diameter d inches.

Guides G are provided for the chain.

Now suppose that the chain is pulled so that the upper pulleys
make one complete revolution. The amount of chain rolling off
on the left, coming from the pulley B, will be πd inches, and
a length of the chain equal to πD inches will roll on on the right,
so that the chain as a whole rolls on a distance equal to ($\pi D - \pi d$)
inches.

The weight will move up half this distance or $\dfrac{\pi (D - d)}{2}$ inches.

The reason for this half requires some further explanation;
we will explain it by considering a rope or chain passing over a
single pulley Q, Fig. 40 (b), and fixed at one end to a point P.
Now suppose that the free end is moved from the position X to
the position X'; the pulley moves up to the position shown in
dotted lines. The rope or chain may be considered as made up
of three lengths; the piece PM on the left; the piece MN

encircling the pulley and the piece NX on the right. In the raised position the pieces are PM', $M'N'$ and $N'X'$.

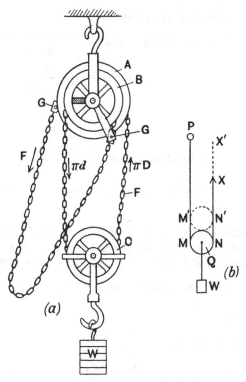

Fig. 40. Weston's Pulley Block.

Therefore total length before movement $= PM + MN + NX$.
Therefore total length after movement $= PM' + M'N' + N'X'$.
Now these two lengths must be the same and clearly

$$MN = M'N';$$
$$\therefore PM + NX = PM' + N'X',$$

i.e. $PM' + M'M + NN' + N'X = PM' + N'X + XX',$
or $M'M + NN' = XX';$
but clearly $MM' = NN' =$ distance moved up by Q,

or $$M'M = \frac{XX'}{2};$$

i.e. distance moved up by pulley $= \frac{1}{2}$ distance moved up by rope or chain.

Summarising our results we have:

$$\text{Distance moved at effort} = \pi D.$$

$$\text{Distance moved at weight} = \frac{\pi D - \pi d}{2};$$

$$\therefore \text{Velocity ratio} = \frac{\pi D}{\frac{\pi D - \pi d}{2}}$$

$$= \frac{2D}{D - d} \quad \cdots\cdots\cdots(1).$$

$$\therefore \text{Ideal effort} = F = \frac{W\,(D - d)}{2D} \quad \cdots\cdots\cdots(2).$$

d is usually made nearly equal to D to make the ideal efforts as small as possible.

Numerical Example. *In a Weston Pulley Block the larger pulley has* 12 *teeth and the smaller has* 11 *teeth. If the efficiency is* 60 *per cent., what load will be raised by an effort of* 20 *lbs.?*

In this case the velocity ratio $\frac{2D}{D - d}$ will be equal to

$$\frac{2 \times 12}{12 - 11} = 24;$$

$$\therefore \text{Ideal effort} = \frac{W}{24},$$

$$\text{Actual effort} = \frac{\text{Ideal effort}}{\text{Efficiency}} = \frac{W}{24} \div \cdot 60;$$

$$\therefore 20 = \frac{W}{24 \times \cdot 60},$$

$$W = 24 \times \cdot 60 \times 20 \text{ lbs.}$$

$$= \underline{288 \text{ lbs.}}$$

Actual Performance of Machines. We have already stated that in practice machines are never ideal and that some energy is always wasted. If we regard the actual effort as the sum of ideal effort and waste effort we shall find that in actual tests the waste effort is almost constant but increases slightly as the load or resistance increases.

The usual procedure in testing a simple machine is to first

find what effort can be exerted, when there is no load on the
machine, before the point at which the effort is applied will move
slowly without increasing in speed. This initial effort is the
initial waste effort and spends itself in lifting the dead weight
of the machine itself and in overcoming the friction or "sticki-
ness" of the various parts.

Various loads are then put on the machine and the effort
necessary to lift each slowly at the same speed is noted carefully.

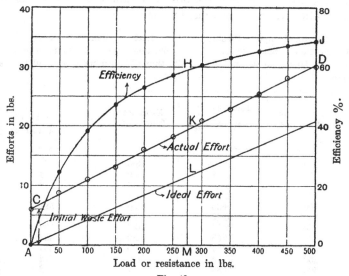

Fig. 41.

To minimise errors it is a good plan to increase the loads gradually
in, say, 10 steps, i.e. 50 lbs. at a time for a maximum load of
500 lbs., until the maximum load is reached and then decrease
the loads gradually by the same amount, the mean of the actual
efforts "ascending" and "descending" being taken as the final
values. The mean actual efforts are then plotted against the
loads as indicated in Fig. 41 and the resulting curve will usually
give a straight line CD.

On the same base are then plotted the values of the Ideal
Efforts. This will give a straight line AG, given by the relation
$AG = \dfrac{AB}{v_r}$, where v_r is the velocity ratio.

Now since Efficiency $= \dfrac{\text{Ideal effort}}{\text{Actual effort}}$ we can obtain values from which we can plot to a convenient scale the efficiency curve AHJ.

For any load, say AL, we have

$$e = \text{Efficiency} = \frac{ML}{KL}.$$

It is preferable to take the values of the actual effort from the line CD, instead of from the observed values, because errors of observation are smoothed out by the curve.

Experiment upon Weston Pulley Block. Take for example the Weston Pulley Block described on p. 69, and suppose that the loads are increased 50 lbs. at a time; the results thus obtained may be tabulated as follows:

Resistance or Load $= W$ lbs.	Ideal Effort $= F_i$ lbs. $= \dfrac{W}{24}$	Mean Actual Effort $= F$ lbs.	Efficiency $= \dfrac{\text{Ideal Effort}}{\text{Actual Effort}}$
0	0	6·0	0
50	2·08	8·9	·234
100	4·17	11·0	·379
150	6·25	13·1	·477
200	8·33	16·0	·521
250	10·42	18·1	·576
300	12·50	20·9	·598
350	14·58	22·9	·637
400	16·67	25·5	·654
450	18·75	28·2	·672
500	20·83	30·1	·692

As an illustration of the suggestion of taking the values of the actual effort, for calculating the efficiency, from the curve CD instead of from the actual observed values we will take the case of $W = 350$ lbs. The observed value of F is 22·9 lbs. but the value obtained from the diagram is 23·1 lbs.; this value was used in calculating the efficiency.

Experiments upon a bicycle gear. The following tests upon a bicycle provided with a two-speed gear were made in order to find the efficiency of the chain drive when driving " solid," and when driving through the speed gear, and can be made in a similar manner upon any bicycle. In order to make the nature of the experiment more clear we will first give a brief explanation of the action of such two-speed gears. Referring to Fig. 42 which shows diagrammatically the arrangement commonly adopted, the chain wheel of the back wheel of the bicycle is connected to an internally toothed wheel B with which engage a number of toothed wheels of pinions C—usually called " planet pinions "—which are carried by a cage D fixed to the hub of the back wheel. The planet pinions C engage also a wheel A—usually called the sun-wheel—which is capable of being fixed to the frame of the bicycle or of rotating freely. In the normal gear, the wheel A is free to rotate and a clutch or mechanical locking device locks the ring B to the cage D thus giving what is called a " solid drive." To put in the low gear the clutch between the ring and cage is released and the wheel A is fixed. The pinions C carried by cage D then have to roll simultaneously upon the fixed pinion A and the ring B and the cage is thus forced to go more slowly than the ring; this means that to drive the wheel, which is connected to the cage D, at a given speed the ring B and therefore the pedals must be rotated more quickly. In other

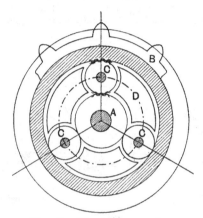

Fig. 42.   Bicycle two-speed gear.

words the gear is lowered. Although we do not intend to explain the derivation of the formula at the present stage, the reduced gear can be calculated as follows :

Let N_A = the number of teeth on the wheel A,
   N_B = the number of teeth on the annular wheel B.

Then   Reduced gear $= \dfrac{N_B}{N_A + N_B} \times$ Normal gear.

Now the " normal gear " of a bicycle is the diameter in inches of the equivalent direct driven single wheel. It is obtained by the rule :

Normal gear in inches

$$= \frac{\text{Diameter of back wheel in inches} \times \text{Teeth on front chain wheel}}{\text{Teeth on small chain wheel}}$$

In the case of the bicycle under consideration we get

$$\text{Normal gear in inches} = \frac{28 \times 52}{18} = 80 \cdot 89.$$

We also have $N_A = 27$ and $N_B = 69$.

$$\therefore \text{ Low gear} = \frac{80 \cdot 89 \times 69}{69 + 27} = \frac{80 \cdot 89 \times 69}{96} = 58 \cdot 1 \text{ inches.}$$

Method of testing. To test the efficiency of the bicycle gearing the bicycle is suspended in the manner indicated in Fig. 43 and a scale pan *B* is connected by a string to the tyre. The cranks are placed in horizontal position and a scale pan *A* is suspended from one of the pedals so that when loaded it will tend to lift the scale pan *B*. In order to make the test as accurate as possible care must be taken that the back wheel is brought by the aid of the free-wheel clutch to the " balanced " position before the scale pan *B* is fixed to it. The tyre-valve is the principal cause of the lack of balance of a bicycle wheel and if a bicycle be lifted off the ground the wheel will start swinging due to the lack of balance and the position of the wheel at which it ultimately comes to rest is the one that is referred to above as the " balanced " position.

In making the test, the scale pan *B* is weighted and placed upon its stop *D* and the scale pan *A* is then loaded carefully until it begins to fall slowly on

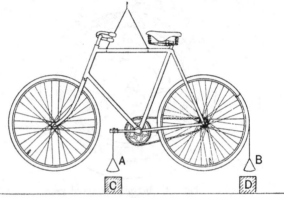

Fig. 43.

to its stop *C*; the load is then carefully taken off until it begins to rise slowly and the scale pan *B* falls on to its stop *D*.

The following results were obtained in an actual test.

Weight acting on tyre (including scale pan) $= 2 \cdot 08$ lbs.

Normal gear:
    Weight acting on pedal to lift scale pan $B = 12 \cdot 25$ lbs.
    Weight acting on pedal to allow scale pan *B* to lower $= 11 \cdot 75$ lbs.

Low gear:
    Weight acting on pedal to lift scale pan $B = 9 \cdot 35$ lbs.
    Weight acting on pedal to allow scale pan *B* to lower $= 7 \cdot 75$ lbs.

We will now work out the velocity ratio.

    The cranks are 7 inches long.

In one revolution of the crank, the distance moved by the centre of the pedal $= 2\pi \times 7$ inches. The wheel moves through a distance $= \pi \times$ gear.

$$\therefore \text{ Normal gear velocity ratio} = \frac{\pi \times 14}{\pi \times 80 \cdot 89} = \frac{1}{5 \cdot 78},$$

$$\text{Low gear velocity ratio} = \frac{\pi \times 14}{\pi \times 58 \cdot 1} = \frac{1}{4 \cdot 15}.$$

Normal gear efficiencies.

Ideal effort to lift scale pan B = resistance × velocity ratio = $2\cdot08 \times 5\cdot78$

$$= 12\cdot02 \text{ lbs.}$$

$$\therefore \text{ Efficiency} = \frac{\text{Ideal effort}}{\text{Actual effort}} = \frac{12\cdot02}{12\cdot25}$$

$$= \cdot980 \text{ or } 98\cdot0 \text{ per cent.}$$

Ideal resistance lifted by scale pan $B = \dfrac{11\cdot75}{5\cdot78} = 2\cdot03\,;$

$$\therefore \text{ Reversed efficiency} = \frac{2\cdot03}{2\cdot08} = \cdot974$$

$$= 97\cdot4 \text{ per cent.}$$

Low gear efficiencies.

Ideal effort to lift scale pan $B = 2\cdot08 \times 4\cdot15$

$$= 8\cdot63 \text{ lbs.}$$

$$\therefore \text{ Efficiency} = \frac{\text{Ideal effort}}{\text{Actual effort}} = \frac{8\cdot63}{9\cdot35}$$

$$= \cdot923$$

$$= 92\cdot3 \text{ per cent.}$$

Ideal resistance lifted by scale pan $B = \dfrac{7\cdot75}{4\cdot16} = 1\cdot867\,;$

$$\therefore \text{ Reversed efficiency} = \frac{1\cdot867}{2\cdot08} = \cdot897$$

$$= 89\cdot7 \text{ per cent.}$$

From the above we see that the efficiency at normal gear is 98·0 and at low gear 92·3; the two-speed gear therefore causes an additional loss of $98\cdot0 - 92\cdot3 = 5\cdot7$ in 98·0 or about $\dfrac{5\cdot7}{98\cdot0} \times 100 = 5\cdot8$ per cent.

Work done on rotating bodies. In the cases that we have considered up to the present we have dealt only with bodies which are moved in a straight line. In a larger number of cases in engineering practice, however, we have to deal with rotating bodies. Take for example the case of a pulley A, Fig. 44, which is being rotated by a belt or chain B. There is a tension T_1 lbs. on the tight side of the belt and a tension T_2 lbs. on the slack side.

Suppose that the pulley makes one revolution and that the belt does not slip and that the radius to the centre of the belt is r feet; the circumference of the pulley then moves the same distance as the belt, i.e. a distance equal to $2\pi r$ in the direction of the effort and resistance.

The work done on the pulley by the tension T_1

$$= 2\pi r T_1.$$

Work taken from the pulley by the tension T_2

$$= 2\pi r T_2;$$

therefore resulting work done on the pulley in one revolution

$$= 2\pi r T_1 - 2\pi r T_2$$
$$= E = 2\pi r (T_1 - T_2) \dots\dots\dots\dots(1).$$

This represents the work done on the machine which the pulley drives. As a rough approximation we may take the tension on the tight side equal to twice that on the slack side.

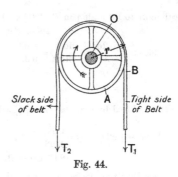

Fig. 44.

Some people derive this result as follows:

Take moments about the centre O of the shaft upon which the pulley is mounted. Then resultant moment

$$= T_1 r - T_2 r = (T_1 - T_2) r.$$

This resultant moment is called the *torque*. This gives rise to the following general rule:

Work done per revolution in ft.-lbs. $= 2\pi \times$ *torque in lbs.-ft.*. . (2).

Now suppose that the pulley makes N revolutions in one minute.

Then the work done on the pulley in one minute

$$= EN = 2\pi r N (T_1 - T_2);$$
$$\therefore \text{Horse-Power} = \frac{\text{Work per min.}}{33,000} = \frac{2\pi r N (T_1 - T_2)}{33,000} \quad . . (3).$$

Numerical Example. *What horse-power is transmitted to a pulley rotating at a speed of 120 revolutions per minute if the*

tension on the tight side is 150 *lbs. and on the slack side is* 75 *lbs.,*
the diameter of the pulley being 18 *inches ?*

In this case $T_1 - T_2 = 150 - 75 = 75$ lbs.;

$$r = \tfrac{9}{12} = \cdot 75 \text{ foot};$$

∴ Work done per minute $= 2\pi \times \cdot 75 \times 120 \times 75$;

$$\therefore \text{Horse-Power} = \frac{2\pi \times \cdot 75 \times 120 \times 75}{33,000}$$

$$= 1\cdot 29.$$

Indicated and Brake Horse-Power of Engines. In testing
steam, gas, oil and similar engines it is usual to measure what

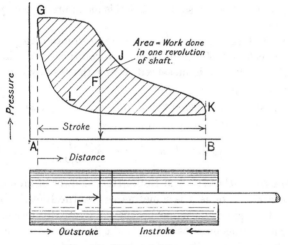

Fig. 45. Indicated Horse-Power.

are called the "Indicated Horse-Power" (I.H.P.) and the "Brake
Horse-Power" (B.H.P.).

The *indicated horse-power* is in a sense a measure of the power
input and is calculated from diagrams drawn by an instrument
called the "indicator" which automatically indicates graphically
as a diagram the pressure of the steam or gas in the engine
cylinder at the various points of the stroke. This diagram in
the case of a steam-engine is somewhat as indicated in Fig. 45.
The total pressure acting upon the piston is the effort so that the
indicator diagram draws for us the effort curve and we have

shown already that the area under the effort curve represents
the work done.

On the outstroke of the piston the work done is represented
by the area $AGJKB$ and on the instroke the work taken from the
piston in bringing it back is represented by the area $BKLGA$;
the difference between these two areas, that is the area shaded,
represents therefore the work done by the steam or gas upon the
piston in one double-stroke of the latter, i.e. in one revolution of
the engine shaft. Therefore the mean height of this diagram,
i.e. $\dfrac{\text{shaded area}}{BA}$, represents the mean effort. The indicator is
calibrated so that by multiplying the mean height of the diagram
in inches by a constant we get at once the mean pressure acting
on the piston in lbs. per sq. in. Let this mean pressure be p_m lbs.
per sq. in.

Now let A be the area of the piston in square inches; L the
stroke in feet and N the number of revolutions per minute of
the engine shaft. N is measured during the test by a counter.
Then $E_m = p_m A$.

$\therefore$ Work done per revolution $= E_m \cdot L = p_m AL$.

$\therefore$ Work done per minute $= p_m ALN$.

$\therefore$ Indicated Horse-Power $= \dfrac{p_m ALN}{33,000}$ ......(1).

Brake Horse-Power. The Brake Horse-Power of an engine is
the power output or as it is sometimes called the Effective Horse-
Power. It is given its name because it is usually measured in
tests by an arrangement called a "brake," a simple form of which
is as follows. A rope B, Fig. 46, is passed over the flywheel A; it
is usually made up of three or four pieces of rope knotted together
at the ends and held apart by distance pieces D. On the side on
which the flywheel would tend to lift it is hung a weight pan C
which is often provided at the bottom with a piece of rope secured
to the floor to prevent the weights from being bodily carried
right over the flywheel. The rope is connected at the other end
to a spring balance the reading of which may vary slightly from
time to time.

As the flywheel rotates in the direction of the arrow, the rope
will slip continuously. We have here exactly the converse of
the belt drive of a pulley. Here the pulley is the driving member,

and it spends its energy in overcoming the friction or grip between the rope and the flywheel. Now let the weight on the pan be W lbs. and the reading of the spring balance w lbs., and let r feet be the radius of the flywheel.

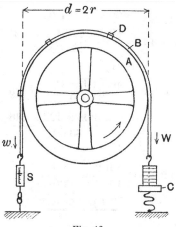

Fig. 46.

In one revolution the flywheel does an amount of work equal to

$$E = 2\pi r W - 2\pi r w = 2\pi r (W - w) = \pi d (W - w).$$

If therefore the flywheel makes N revolutions per minute we have

Work per minute output of engine $= EN = \pi dN (W - w)$;

$$\therefore \text{Brake Horse-Power} = \frac{EN}{33,000},$$

i.e. B.H.P. $= \dfrac{\pi dN (W - w)}{33,000}$ ....(2).

The diameter d should be measured to the centre of the rope.

The ratio $\dfrac{\text{B.H.P.}}{\text{I.H.P.}}$ is called the *mechanical efficiency* of the engine.

Numerical Example. *In the test of a steam-engine the mean pressure was found from the indicator diagram to be 60·3 lbs. per sq. in. and the stroke was 12 inches. The piston was of 10 inches diameter and the number of revolutions per minute was 122. The*

*diameter of the flywheel was 5 feet and the rope was 1 inch in diameter,
and the weight W was 240 lbs., the spring balance reading being 4 lbs.
Find the I.H.P., B.H.P. and mechanical efficiency.*

$$\text{Area of piston} = A = \frac{\pi}{4} \times 10^2 = 78{\cdot}54 \text{ sq. ins.,}$$

$$p_m = 60{\cdot}3, \quad L = 1 \text{ ft.} ;$$

$$\therefore \text{I.H.P.} = \frac{60{\cdot}3 \times 78{\cdot}54 \times 1 \times 122}{33,000}$$

$$= 17{\cdot}5,$$

$$\text{B.H.P.} = \frac{\pi d N\,(W - w)}{33,000}, \quad d = 5 + \frac{1}{12} = 5{\cdot}08 \text{ ft.}$$

$$= \frac{\pi \times 5{\cdot}08 \times 122 \times 236}{33,000}$$

$$= 13{\cdot}9,$$

$$\text{Mechanical efficiency} = \frac{\text{B.H.P.}}{\text{I.H.P.}} = \frac{13{\cdot}9}{17{\cdot}5} = {\cdot}794$$

$$= \underline{79{\cdot}4 \text{ per cent.}}$$

SUMMARY OF CHAPTER IV.

A *machine* is an appliance for receiving energy from some outside source and converting it into some more convenient form.

$$\text{Mechanical advantage} = \frac{\text{Resistance}}{\text{Effort}}.$$

$$\text{Velocity ratio} = \frac{\text{Distance moved at the effort in a given time}}{\text{Distance moved at the resistance in a given time}}.$$

The *efficiency* of a machine is the ratio of the energy that it gives out to the energy that it receives.

$$\text{Efficiency} = \frac{\text{Ideal effort}}{\text{Actual effort}}.$$

Mechanical advantage = velocity ratio × efficiency.

If a machine is " self-sustaining " or not reversible, its efficiency cannot be as much as 50 per cent.

Work done upon rotating bodies per revolution in ft.-lbs. = 2π × torque in lb.-ft.

Indicated Horse-Power (I.H.P.) of an engine $= \dfrac{p_m A L N}{33,000}$,

Brake Horse-Power (B.H.P.) of an engine $= \dfrac{\pi d N (W - w)}{33,000}$,

Mechanical efficiency $= \dfrac{\text{B.H.P.}}{\text{I.H.P.}}$.

EXERCISES. IV.

1. In a wheel and axle the diameter of the wheel is 3 ft. 6 ins. and the diameter of the axle is 10 ins. The diameter of the rope attached is in each case 1 in. Find the weight which can be lifted by a pull of 50 lbs. on the rope attached to the wheel.

2. If in the last example a weight of 195 lbs. is lifted what is the efficiency of the machine?

3. The diameter of the wheel in a wheel and axle is 18 ins., and that of the axle 5 ins. Neglecting friction what pull on the wheel will raise a weight of 600 lbs.? If it requires a pull of 200 lbs. weight to lift this load what is the efficiency? Also find the mechanical advantage and velocity ratio of the machine.

4. Find the H.P. of an engine which will raise 1000 gallons of water per min. from a depth of 240 ft. The efficiency of the engine is 55 per cent.

5. The inclination of a plane is 3 in 5. Find what force acting parallel to the plane will support a load of 2 tons neglecting friction. Also find the force which would be required acting parallel to the base of the plane.

6. The handle of a lifting jack measures 24 ins. in length and the pitch of the screw is $\frac{3}{4}$ in. What force applied at the end of the handle would be required to raise a load of 22 cwts., the effect of friction being neglected?

7. A shaft transmits 50 H.P. at 250 revs. per min. Find the twisting moment in inch-lbs.

8. The twisting moment on an engine shaft is 20,000 in.-lbs. and it makes 180 revolutions per min. Find the H.P. transmitted.

9. The pitch on a screw-jack is $\frac{1}{2}$ inch, the distance from the axis of the screw to the end of the handle 26 inches. Find the velocity ratio. If the law is $F = \cdot 03W + 9\cdot 45$, find the load which will be lifted by a force of 56 lbs. wt. applied at the end of the handle. Find also the efficiency at this load.

10. The diameter of a steam-engine cylinder is 9 ins., the length of crank 9 ins., the number of revolutions per min. 110, and mean effective pressure of the steam 35 lbs. per sq. in.; find the indicated H.P.

11. In measuring the brake H.P. of an engine a rope passes round the flywheel, one end being fixed to a spring balance; the other end carries a weight of 120 lbs. If the wheels make 150 revs. per min. and the spring balance indicates 15 lbs. what is the H.P. transmitted? The flywheel is 5 ft. in diameter.

12. A steam pump is to deliver 1000 gallons of water per minute against a pressure of 100 lbs. per sq. in. Taking the efficiency of the pump to be ·70, what indicated H.P. must be provided?

13. The diameter of the cylinder of a double acting engine is 10″, stroke 15″, number of revolutions per min. 120, and the mean steam pressure 48 lbs. per sq. in. Find the H.P. transmitted.

14. In a rope-brake dynamometer the diameter of the brake wheel is 10 ft., rope is $1\frac{1}{2}''$ diameter, weight on rope at one end is 200 lbs. and pull on spring balance at the other end is 18 lbs. weight. If the wheel makes 90 revs. per min. find the H.P. transmitted.

15. The following results were obtained in a test of a steam-engine: 1 H.P. = 7·7; revs. per min. = 164; diameter of brake-wheel 3 ft.; diameter of rope $\frac{1}{2}$ in.; weight on brake 150 lbs.; reading of spring balance 2·5 lbs. Find the mechanical efficiency of the engine.

16. The following results were obtained in a test of a machine whose velocity ratio = 8:

Load or resistance lbs.	0	5	10	20	30	40
Effort lbs.	3·0	4·8	6·0	10·0	14·0	18·0

Plot a curve showing the efficiencies at various loads and find the efficiency for a load of 25 lbs.

CHAPTER V

VELOCITY AND ACCELERATION

WHAT do we mean when we say that a train is going at 60 miles an hour at a certain point? We do not mean that in one hour the train actually goes 60 miles; it might stop altogether after it has gone 20 miles. But what we mean is that if the train continued to move at the same speed or velocity for one hour it would then have gone 60 miles. Expressing velocity in scientific language we say that "velocity or speed is the rate of change of position or space with respect to time."

Velocity is a vector quantity*; its direction is of importance as well as its magnitude. It is here that many people use the term "speed" and "velocity" with a slight difference of meaning. When speed is spoken of the direction does not come into consideration but velocity involves the direction of the motion.

Velocity of a point. When a body is moving, different points in it may be moving with different velocities, so that in strict language we do not speak of the velocity of a body but of the velocity of a point.

Uniform Velocity. The velocity of a point is said to be uniform when it maintains the same direction and magnitude (i.e. the point passes through equal distances in the same direction in equal times).

Suppose that a point has a uniform velocity of 10 feet per second in a certain direction. Then in 1 second it will move through 10 feet; in 100 seconds it will move through 1000 feet; in $\frac{1}{100}$ second it will move through $\frac{1}{10}$th of a foot and so on. This is expressed in symbols as follows: If a point has a uniform

* Cf. p. 1.

velocity v feet per second, then the distance s in feet covered in t seconds is given by the formula

$$s = vt \dots\dots\dots\dots\dots\dots\dots(1).$$

This is true no matter how large or small t may be.

Variable Velocity. In practice velocity is seldom if ever uniform although it may over a certain time be sufficiently nearly so to be reckoned as uniform for all practical purposes.

There are two causes that may disturb uniformity of velocity:

        (a)   Variation in magnitude.

        (b)   Variation in direction.

Very often these two variations occur together, but for the rest of this chapter we will consider change in magnitude only.

Velocity variable in magnitude. Suppose that the times are recorded at which a moving point passes certain stations and that the distances of these stations from a suitable starting point are plotted against the times of passing. The curve $CPQD$, Fig. 47, obtained by joining up the points is called the *space curve*.

At the instant from which the time is reckoned the distance from the starting point is AC; then at any point such as P, after a time $t = AT$ has elapsed, the moving point is at a distance $s = PT$ from its starting point.

Now consider a point Q on the space curve very near to P, and let PR be drawn perpendicular to QU; while the point has moved a distance equal to $QU - PT = QR$, the time has increased by an amount $TU = PR$.

Now $\dfrac{QR}{PR} = \dfrac{\text{Distance moved}}{\text{Time taken}} = \tan \theta.$

Next suppose that the points Q and P move closer and closer to each other; the line PQ then gradually approaches the position of the tangent XY shown in dotted lines, and the slope of this tangent may be taken as $\tan \theta$ if PQ is sufficiently small.

Now we define the velocity at any point as the value which the $\dfrac{\text{distance moved}}{\text{time taken}}$ approaches as the distance moved becomes smaller and smaller. It follows from this that *the slope of the tangent to the space curve at any point measures the velocity at that point.*

In working from the diagram we must be careful to allow properly for the scales; referring to the figure we have

$$\text{Velocity at given point} = \tan \theta = \frac{YZ \text{ on space scale}}{XZ \text{ on time scale}}.$$

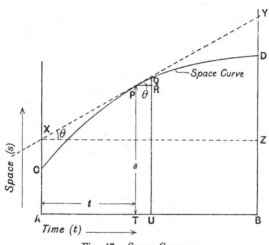

Fig. 47.   Space Curve.

Numerical Example. *The following results were obtained in timing a man walking over a certain distance. Find the velocity at the commencement and after 40 minutes from the start of the test. Find also the average velocity over the whole test.*

Time in seconds	0	10	20	30	40	50	60
Distance in feet from starting point	150	196	263	345	440	505	550

The results are plotted in Fig. 48 where *BCD* is the space curve and we are required to find the velocities at *B* and *C*. To do this we have, as previously explained, to draw tangents to the space curve at *B* and *C*. This is not easy to do accurately; so far no graphical construction is known which is more accurate than drawing a line by eye to touch the curve.

Then Velocity at $B = \tan\theta_B = \dfrac{EK}{BE} = \dfrac{241 \text{ feet}}{60 \text{ seconds}}$

$$= 4\cdot01 \text{ feet per second.}$$

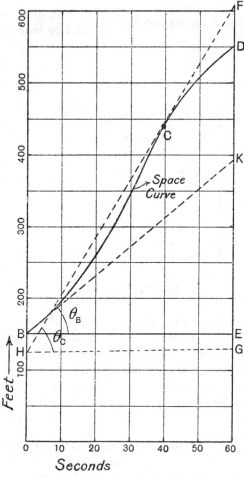

Fig. 48.

It is common to measure some velocities in miles per hour.
Now 1 mile per hour = 5280 feet in 3600 seconds

$$= \frac{5280}{3600} = \frac{22}{15} = 1\cdot467 \text{ feet per second;}$$

$\therefore$ Velocity at B in miles per hour $= \dfrac{4\cdot01}{1\cdot467} = 2\cdot73$ miles per hour,

Velocity at $C = \tan\theta_c = \dfrac{GF}{HG} = \dfrac{480}{60} = 8$ feet per second

$$= \dfrac{8}{1\cdot467} = 5\cdot45 \text{ miles per hour.}$$

Average velocity $= \dfrac{DE}{BE} = \dfrac{400}{60} = 6\cdot67$ feet per second

$$= \dfrac{6\cdot67}{1\cdot467} = 4\cdot55 \text{ miles per hour.}$$

A useful figure to remember is that 60 miles an hour is equal to 88 feet per second, or one mile per hour = $\frac{22}{15}$ feet per second.

Velocity Curve and its relation to the Space Curve. Next suppose that we know the velocity at every time and that we plot velocities upon a time base; then the resulting curve is called a velocity curve $GHJK$, Fig. 49. AG represents the velocity at the beginning of the period of time under consideration and is called the *initial velocity* and will be given the letter u.

Now consider the relation between this curve and the space curve plotted on the same base (to save confusion it is preferable to plot one diagram above or below the other). We have already shown that the velocity v at the middle of a very short time LM is equal to the slope of the tangent to the space curve at the corresponding point. Since LM is so very short, the slope of this tangent is given by

$$\tan\theta = \dfrac{QR}{PR} = \dfrac{QR}{LM};$$

$$\therefore v = \dfrac{QR}{LM};$$

$\therefore QR = LM \times v =$ Area of shaded strip of the velocity curve. But QR is the increase in the ordinate of the space curve, and we could show similarly for any other strip that the increase in the ordinate of the space curve represents the area of the corresponding strip of the velocity curve.

$\therefore QU =$ Total increase in space from the beginning = Area of velocity curve from A to M, i.e. area $AGJM$. But this is exactly the relation which we have explained (p. 43) between a slope curve and its primitive curve.

Therefore the space curve is the sum curve of the velocity curve.

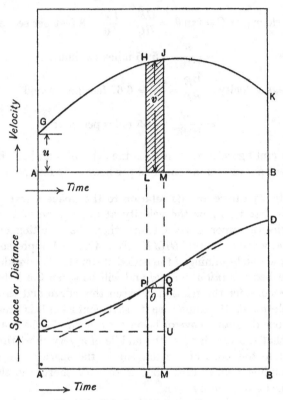

Fig. 49. Velocity and Space Curves.

We may use as a general rule the following relation which we have proved above. If a curve A is the sum curve of a curve B, the ordinate of B at any point represents the slope of A at the corresponding point.

Some special cases of Velocity and Space Curves. (*a*) *Constant velocity.* If the velocity of a point is constant, the velocity curve is a horizontal straight line, Fig. 50. If the sum curve construction be carried out for this we get a sloping straight line AD, assuming that we commence reckoning our distances from the point at which the time commences. This is because all the mid-ordinates of the velocity curve when projected horizontally

come to the point G so that all the elemental pieces of the sum
curve are parallel to PG. [If, instead of taking our distances from
the point at which the time commences, we take them from some
other point, we shall get a space curve such as cR parallel to
AD, but in all further cases we shall assume that the space is
considered as commencing at the beginning of the time interval.]

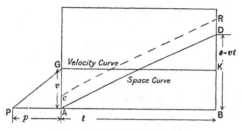

Fig. 50. Constant Velocity.

We then have: distance covered from A to B

$$= \text{area } AGKB = vt,$$

i.e. $$s = vt \dots\dots\dots\dots(1).$$

Since for any value of t the space s is equal to vt, the space
curve is a straight line such that $\tan DAB = \dfrac{s}{t} = v$.

(b) *Velocity increasing uniformly.* If the velocity increases
by the same amount in each unit of time we shall obtain for our
velocity curve a sloping straight line GK, Fig. 51.

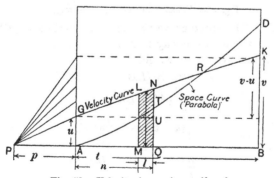

Fig. 51. Velocity increasing uniformly.

If we apply to this the sum curve construction the space
curve will be found to be a *parabola ARD*.

We then have: distance covered from A to B

$$= \text{area } AGKB = \tfrac{1}{2} AB\, (AG + BK)$$

$$= \tfrac{1}{2}t\,(u + v),$$

i.e.

$$s = t\left(\frac{u + v}{2}\right)\dots\dots\dots\dots\dots(2).$$

We shall see later that it is sometimes more convenient to write this

$$s = \frac{t}{2}\{u + (u + v - u)\}$$

$$= \frac{t}{2}\{2u + (v - u)\}$$

$$= ut + \frac{(v - u)\,t}{2}\ \dots\dots\dots\dots\dots(3).$$

(c) *Velocity decreasing uniformly.* In this case the velocity curve GK, Fig. 52, will also be a straight line but will slope

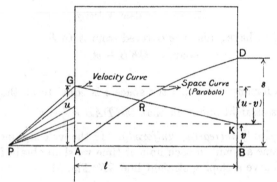

Fig. 52. Velocity decreasing uniformly.

downwards. The space curve will also be a parabola ARD, but it will curve the opposite way from the previous case.

After a time t therefore we get

$$s = \text{area } AGKB$$

$$= \frac{AB}{2}\,(AG + BK)$$

$$= \frac{t}{2}\,(u + v) \text{ as before}$$

$$= \frac{t}{2}\{u + u - (u - v)\}$$

$$= \frac{t}{2}\{2u - (u - v)\}$$

$$= ut - \frac{(u - v)}{2}\,t\ \dots\dots\dots\dots\dots(4).$$

Numerical Example. *A point starts from rest and increases its velocity uniformly for* 10 *seconds at the end of which it has a velocity of* 10 *feet per second. It continues to move for* 10 *more seconds at this velocity and the velocity then diminishes uniformly for* 5 *seconds when it comes to rest. How far has it travelled ?*

The velocity curve for this case is as shown in Fig. 53. For

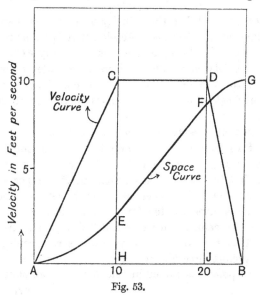

Fig. 53.

the first 10 seconds between *A* and *C* it is a sloping straight line *AC*; for the next 10 seconds the velocity is constant so that the velocity curve is a horizontal straight line *CD* and for the next 5 seconds the velocity falls uniformly to zero so that the velocity diagram is the sloping straight line *DB*.

Now the total space covered in the 25 seconds will be represented by the area *ACDB*.

This can be estimated as follows :

$$\text{Area of } \triangle ACH = \frac{10 \times 10}{2} = \ 50 \text{ feet,}$$

$$\text{Area of rectangle } CDJH = 10 \times 10 = 100 \text{ feet,}$$

$$\text{Area of } \triangle DJB = \frac{10 \times 5}{2} = \ 25 \text{ feet,}$$

$$\text{Total distance covered} = \underline{175} \text{ feet.}$$

The space curve will be as indicated, AE and FG being parabolic arcs and EF a straight line. As an exercise the reader should draw the curve by the sum curve construction.

Suitable scales would be as follows: time $1'' = 5$ seconds; velocity $1'' = 5$ feet per second; polar distance $= 2$ inches.

Then the space scale will be $1'' = 2 \times 5 \times 5 = 50$ feet so that BG should measure $3 \cdot 5$ inches.

Acceleration. When the velocity of a body is changing it is said to have an *acceleration*. Acceleration is measured by the rate of change of velocity and we have already explained that change of velocity may take place in magnitude or direction or both. For the present we will confine ourselves to change of magnitude of velocity and will assume that the direction remains constant.

Suppose that a body is moving at a certain instant with a velocity of 10 feet per second and that one second later it is moving in the same direction with a velocity of 12 feet per second. In one second the velocity has gained by 2 feet per second, so we say that the mean acceleration is 2 feet per second per second; this is often written for brevity 2 ft./sec.[2]

Now let the velocity curve be $LMNK$, Fig. 54. At the time represented by the point T the velocity is represented by TM and after a short time TU it is represented by UN, so that in time TU the point has gained in velocity by an amount NO.

$$\therefore \text{ Mean velocity gained in unit time} = \frac{NO}{MO}.$$

Now the points TU are very close together and as N comes closer still to M the line joining MN ultimately becomes the tangent XX to the velocity curve at the point M.

Then rate of change of velocity

$$= \frac{NO}{MO} = \text{slope of tangent } XX = \tan \theta.$$

Therefore the acceleration at any point is represented by the slope of the velocity curve at the given point.

If we obtain the accelerations at a number of points and plot them against the times, the resulting curve will be an acceleration curve.

Positive and negative acceleration. We have up to the present

only spoken of velocity gained but the term "gain" must be
considered as including "loss" and when there is a loss we shall
regard it as a negative gain. Returning to our numerical
illustration suppose that instead of being 12 feet per second at
the end of one second the velocity is 8 feet per second; in one
second the velocity has lost 2 feet per second and we should say
that the mean acceleration is − 2 feet per second per second.

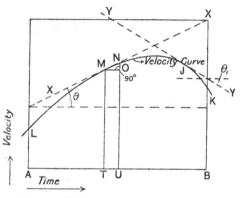

Fig. 54. Acceleration.

Now when the velocity is decreasing the tangent, such as YY,
Fig. 54, cuts the base at a point in advance of the point of contact
whereas when the velocity is increasing the tangent cuts the
base behind the point of contact.   This enables us to formulate
the following rule.   If the tangent to the velocity curve cuts the
time base at a point behind the point of contact, the acceleration
is positive and if it cuts at a point beyond the point of contact
the acceleration is negative.

General relation between Acceleration, Velocity and Space
Curves. We have shown that the slope at any point of the
velocity curve determines the acceleration and we have previously
shown that the slope at any point of the space curve gives the
velocity; there is therefore the same relation between the
acceleration and velocity curves as there is between the velocity
and the space curves.   We get therefore the following very
important rule.

*The velocity curve is the sum curve of the acceleration curve and
the space curve is the sum curve of the velocity curve.*

This is illustrated in Fig. 55 in which, to save confusion, the three curves have been drawn upon separate bases. CDE is the acceleration curve; drawing a sum curve with polar distance p_1, we get the velocity curve AFG if the point starts from rest; if the point has an initial velocity u we set up $AA' = u$ on the velocity scale, obtained as described later, and start the sum curve at

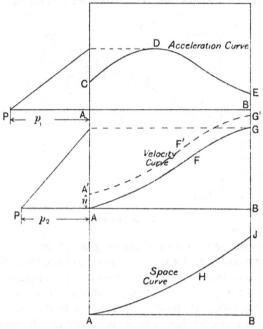

Fig. 55. Relation of Acceleration, Velocity and Space Curves.

A' thus obtaining the velocity curve $A'F'G'$ shown in dotted lines. Drawing the sum curve of this with a polar distance p_2 we get the space curve AHJ.

Scales. Suppose that the time scale is $1'' = x$ seconds, and that the acceleration scale is $1'' = y$ ft./sec.^{2} and suppose that p_1 is measured in actual inches. Then the velocity scale will be $1'' = p_1 xy$ ft./sec. Now let p_2 be also measured in actual inches. Then the space scale will be

$$1'' = p_2 p_1 x^2 y \text{ feet.}$$

This may be explained as follows.

One square inch of the acceleration curve represents xy units of velocity and the sum curve construction gives the area divided by the polar distance so that one inch on the sum curve AFG represents $p_1xy = z$ say.

By similar reasoning one inch on the sum curve AHJ represents $p_2xz = p_1p_2x^2y$.

As a numerical illustration let the time scale be $1'' = 10$ seconds and the acceleration scale $1'' = 2$ feet per second per second and let $p_1 = 2$ inches; then the velocity scale will be

$$1'' = 2 \times 2 \times 10 = 40 \text{ feet per second.}$$

Next let $p_2 = 1\frac{1}{2}$ inches; then the space scale will be

$$1'' = 1\frac{1}{2} \times 40 \times 10 = 600 \text{ feet.}$$

By a careful choice of the polar distances p_1, p_2 we can obtain convenient scales; we should, for instance, have done better in the above case to have taken $p_1 = 2\frac{1}{2}$ inches and $p_2 = 2$ inches, our velocity scale would then be $1'' = 50$ feet per second and the space scale $1'' = 1000$ feet.

Constant Acceleration; equations of motion. If the acceleration is constant, the acceleration curve is a horizontal straight

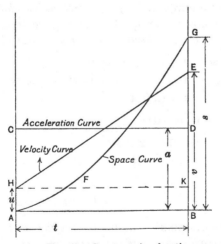

Fig. 56.　Constant Acceleration.

line CD, Fig. 56; the sum curve, i.e. the velocity curve of this, will be the sloping straight line HE, while the space curve AFG will be a parabola, the sum curve of a sloping straight line being a parabola.

From these curves we can deduce the following formulae:

$$KE = \text{area of acceleration curve} = at;$$
$$\therefore v = BK + KE$$
$$= u + at \dots\dots\dots\dots\dots\dots\dots\dots(5).$$
$$s = \text{area } AHEB$$
$$= t\frac{(u+v)}{2} = t\left(\frac{u+u+at}{2}\right)$$
$$= ut + \tfrac{1}{2}at^2 \dots\dots\dots\dots\dots\dots(6).$$

We can get a third relation as follows:

By squaring equation (5) we have

$$v^2 = (u^2 + at)^2 = u^2 + 2uat + a^2t^2$$
$$= u^2 + 2a\,(ut + \tfrac{1}{2}at^2)$$
$$= u^2 + 2as \text{ [from (6)]} \dots\dots\dots\dots\dots(7).$$

These equations (5) to (7) are often called the equations of motion and are very useful in problems in which the acceleration is constant.

Numerical Examples. (1) *A point moves along a straight line under an acceleration of 10 ft./sec.*^{2} *The initial velocity is 7 ft./sec. What is the velocity after it has passed over 12 feet?*

In this case $u = 7$ feet per second,
$$a = 10 \text{ feet per second per second,}$$
$$s = 12 \text{ feet.}$$

Therefore using equation (7)

$$v^2 = 7^2 + 2 \times 10 \times 12$$
$$= 49 + 240$$
$$= 289,$$
$$v = \sqrt{289} = \underline{17 \text{ feet per second.}}$$

(2) *A train is running at 20 miles an hour and is stopped by brakes in 10 seconds, the retardation being constant. At how many yards from the stopping point were the brakes applied?*

60 miles an hour = 88 feet per second.

$$\therefore 20 \text{ miles an hour} = \frac{88}{3} \text{ feet per second.}$$

In this case $u = \dfrac{88}{3}$ ft./sec., $t = 10$ and $v = 0$;

$$\therefore 0 = \dfrac{88}{3} + 10a \quad \text{[from (5)]};$$

$$\therefore a = -\dfrac{88}{30} \text{ ft./sec.}^2;$$

$$\therefore s = ut + \tfrac{1}{2}at^2$$

$$= \dfrac{88}{3} \times 10 - \dfrac{88}{2 \cdot 30} \cdot 100$$

$$= \dfrac{880}{3}(1 - \tfrac{1}{2}) = \dfrac{440}{3} \text{ feet} = \dfrac{440}{9} \text{ yards}$$

$$= \underline{48 \cdot 89 \text{ yards.}}$$

Gravity Acceleration " g." If bodies are allowed to drop freely they will be found to have an acceleration which is practically constant.

This acceleration is called the gravity acceleration and is given the letter g. Its value varies slightly with the latitude and with the height above sea-level and in London is usually taken as 32·2 feet per second per second. We will indicate later an interesting simple experiment for determining g and will now derive simplified formulae for the case of bodies falling freely from rest. In equations (5) to (7) therefore we have $u = 0$ and $a = g$ and it is usual to replace the distance or space s by the height h. Our formulae therefore become

$$v = gt \dotfill (8),$$

$$h = \tfrac{1}{2}gt^2 \dotfill (9),$$

$$v^2 = 2gh \dotfill (10).$$

Formula (10) is of the greatest possible importance and may be rewritten in the forms

$$v = \sqrt{2gh} \dotfill (11),$$

$$h = \dfrac{v^2}{2g} \dotfill (12).$$

The student must make himself absolutely familiar with these formulae and should not feel fully satisfied until he can work successfully through all the exercises at the end of the present chapter.

Numerical Example. *A stone is let fall down a well and the splash is heard 2·9 seconds later. If the time for the sound to travel to the top of the well be neglected, what is the depth of the well?*

Let h be the depth of the well.

Then the time in dropping is obtained by equation (9)

$$h = \tfrac{1}{2}gt^2, \text{ i.e. } t = \sqrt{\frac{2h}{g}};$$

but

$$t = 2 \cdot 9 \text{ seconds,}$$

$$\therefore 2 \cdot 9 = \sqrt{\frac{2h}{32 \cdot 2}}.$$

Therefore, squaring,

$$2 \cdot 9^2 = \frac{h}{16 \cdot 1},$$

$$h = 16 \cdot 1 \times 2 \cdot 9^2 = \underline{135 \text{ feet approx.}}$$

With what velocity must a stone be projected if it is to reach a height of 120 feet?

Here we have in our general equation

$$v^2 = u^2 + 2as, \quad v = 0, \quad a = -g, \quad s = h,$$

$$\therefore 0 = u^2 - 2gh,$$

$$u = \sqrt{2gh}$$

$$= \sqrt{2 \times 32 \cdot 2 \times 120} = \underline{88 \text{ ft./sec. approx.}}$$

Limits of use of simple formulae. In using these simple formulae care must be taken to remember that they are based upon the assumption that g is constant and that the bodies fall "freely," i.e. that the air resistance is negligible.

As a matter of fact the air resistance is appreciable for great heights with light bodies; but for this fact a rain drop in falling from a cloud would acquire such a high velocity that it would kill a man if unprotected by armour. Moreover, if the height is very great the body will not fall vertically, judged by standards upon the earth. This point was illustrated in an interesting manner in some experiments which were carried out in a deep vertical mine shaft in the United States of America, one of the shafts being 5300 feet deep.

Smooth metal balls 2 inches in diameter were suspended by threads and allowed to drop by burning the thread, a box of clay being placed 4200 feet beneath. All the balls struck the east wall of the shaft before reaching the box. This was due to the movement of the earth from west to east, this movement being sufficient to cause the balls to be struck by the east wall before they came to the box. In one case 800 feet of fall was sufficient to make a ball dropped 4 feet from the east wall strike against it.

Distance moved in a particular second. In several problems upon velocities and accelerations we require to consider the distance moved in a particular second under a constant velocity.

Suppose for instance that we want to know the distance moved through in the fifth second. In five seconds it will have moved through a certain distance s_5, given by, putting $t = 5$ in equation (6),

$$s_5 = 5u + \frac{25a}{2} \dots\dots\dots(13).$$

In four seconds it will have moved through a distance s_4 given by

$$s_4 = 4u + \frac{16a}{2} \dots\dots\dots(14).$$

Now the difference between the distances moved in five and four seconds respectively must give the distance in the fifth second, so that we have

$$\text{Distance moved in fifth second} = s_5 - s_4$$
$$= u + \frac{9a}{2} \dots\dots(15).$$

Now take the most general case. It is clear from the above illustration, which could be employed for any numerical value, that the distance moved through in the nth second must be the difference between the distances moved through in n and $(n-1)$ seconds respectively,

i.e.   Distance moved through in nth second

$$= s_n - s_{n-1}$$
$$= \{un + \tfrac{1}{2}an^2\} - \{u(n-1) + \tfrac{1}{2}a(n-1)^2\}$$
$$= \{un + \tfrac{1}{2}an^2\} - \{un - u + \tfrac{1}{2}a(n^2 - 2n + 1)\}$$
$$= un + \tfrac{1}{2}an^2 - un + u - \tfrac{1}{2}an^2 + \frac{2an}{2} - \frac{a}{2}$$
$$= u + \frac{2an}{2} - \frac{a}{2}$$
$$= u + \frac{a}{2}(2n-1)$$
$$= u + a(n - \tfrac{1}{2}) \dots\dots\dots\dots(16).$$

Now $(n - \frac{1}{2})$ is the time to the middle of the nth second. Therefore by equation (5) the velocity at that instant

$$= v = u + at$$
$$= u + a\,(n - \tfrac{1}{2}).$$

We thus obtain the very useful rule that: *The space in feet moved through in any particular second is the velocity in feet per second at the middle of that second.*

We have proved this result in the above manner to give us an exercise in reasoning by the manipulation of formulae but we could have proved it, perhaps more simply, from a consideration of the velocity diagram as follows: referring to Fig. 51 let NQ represent the ordinate of the velocity diagram at the end of n seconds and let LM represent it at the end of $(n - 1)$ seconds, so that MQ represents the nth second. Since the space curve is the sum curve of the velocity curve the increase TU in the space over this second is represented by the area of the strip $LNQM$ of the velocity curve and this is equal to $1 \times$ mid-ordinate $=$ velocity at the middle of the second under consideration.

Numerical Example. *A train in two successive seconds moves through* 20·5 *and* 23·5 *feet respectively. If it is being accelerated uniformly what is its acceleration and what was the velocity at the beginning of the first second?*

Suppose that u is the initial velocity.

At the end of the first second we have

$$s = ut + \tfrac{1}{2}at^2,$$
$$20\text{·}5 = u\,.\,1 + \tfrac{1}{2}a\,.\,1 \quad \dots\dots\dots\dots(1).$$

At the end of the second second we have

$$(20\text{·}5 + 23\text{·}5) = u\,.\,2 + \tfrac{1}{2}a\,.\,2^2 \dots\dots\dots\dots(2),$$

$$\therefore 20\text{·}5 = u + \frac{a}{2} \quad \text{from (1)},$$

$$23\text{·}5 = u + \frac{3a}{2} \quad \text{by subtracting (1) from (2)};$$

$$\therefore 3 = a \quad \text{by subtraction};$$

$$\therefore u = 20\text{·}5 - \frac{a}{2} = 20\text{·}5 - 1\text{·}5 = 19;$$

i.e.   Acceleration $= 3$ ft./sec.²,
        Initial velocity $= 19$ ft./sec.

SUMMARY OF CHAPTER V.

Velocity is the rate of change of position with respect to time.

Acceleration is the rate of change of velocity.

The velocity of a point is measured by the slope of the tangent of the space curve.

The acceleration of a point is measured by the slope of the tangent of the velocity curve.

For bodies moving under constant acceleration,

$$v = u + at,$$
$$s = ut + \tfrac{1}{2}at^2,$$
$$v^2 = u^2 + 2as.$$

For bodies starting from rest under gravitational acceleration,

$$h = \frac{v^2}{2g},$$
$$t = \frac{v}{g}.$$

In all problems it is better to reason out as much as possible from first principles than to attempt to remember the formulae and to apply them directly.

EXERCISES. V.

1. What will be the velocity of a body after falling 25 ft. from rest?

2. Find the average speed of a train which runs from London to Grantham, a distance of $105\tfrac{1}{2}$ miles, in 1 hour 55 minutes.

3. A stone takes $2\tfrac{1}{2}$ secs. to drop to the bottom of a well. What is the depth of the well?

4. Suppose a body to have fallen h feet in t secs. from rest according to the law $h = 16{\cdot}1t^2$. Find how far it falls between the times $t = 3$ and $t = 3{\cdot}1$; between $t = 3$ and $t = 3{\cdot}01$; between $t = 3$ and $t = 3{\cdot}001$. Find the average velocity in each of these intervals of time. What do we mean by the actual velocity when t is 3 secs.?

5. What is an acceleration of 60 miles per hour per minute in feet per sec.²?

6. x and t are the distance in miles and the time in hours of a train from a railway terminus:

x	0	1·5	6·0	14·0	19·0	21·0	21·5	21·8	23·0	24·7	26·8
t	0	0·1	0·2	0·3	0·4	0·5	0·6	0·7	0·8	0·9	1·0

Plot on squared paper. Describe why it is that the slope of the curve shows the speed. What is the greatest speed in this case and where approximately does it occur?

7. The following numbers give v the speed of a train in miles per hour at the time t hrs. since leaving a railway station. Draw a diagram showing the distance covered at the various times and find the total distance covered.

v	0	2·4	4·7	7·2	9·6	12·0	14·3
t	·00	0·04	0·08	·12	·16	·20	24·9

v	16·9	18·9	20·7	22·2	23·4	24·3	24·0
t	·28	·32	·36	·40	·44	·48	·52

8. Express 2 ft. per sec. in cms. per min.

9. A train which has constant acceleration starts from rest, and at the end of 3 secs. has a velocity with which it would travel through 1 mile in five mins. Find the acceleration.

10. A train goes from one station to another 5 miles off in 8 mins., first moving with constant acceleration and then with an equal retardation. Find its greatest speed.

11. A train reduces speed from 45 miles an hr. to 15 miles an hr. in 800 yds. How much farther will it go without stopping?

12. A point starting from rest passes over 121 ft. in the sixth second. What is the acceleration?

13. From a balloon which is ascending with a velocity of 32 ft. per sec., a stone is let fall and reaches the ground in 17 secs. How high was the balloon when the stone was dropped?

14. A train goes a distance of 120 miles in 3 hours. During the first hour the speed rises uniformly from rest; during the second hour it remains constant; and during the third hour it falls uniformly to rest. What is the speed during the second hour?

CHAPTER VI

VELOCITY CHANGE IN DIRECTION ; RELATIVE VELOCITY

WE have considered so far only the case of motion in a straight line and have taken into consideration only changes in magnitude of the velocity; but we may also have change in direction, with or without change in magnitude as well. The case of a body moving with constant velocity in a circle is an example in which the magnitude of the velocity is constant but the direction is constantly changing.

Combination of Velocities. The actual velocity possessed by a point may be the combination of two or more velocities, and as velocities are vector quantities, they are added together in

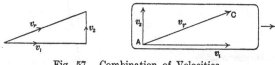

Fig. 57.   Combination of Velocities.

exactly the same way as forces, i.e. by the law of vector addition. Suppose for instance that we are standing at one end A (Fig. 57) of a railway carriage moving with a velocity v_1 and that we walk across the carriage with a velocity v_2; then our actual velocity will be the combination of the velocity v_1 of the train itself and of our own velocity v_2, i.e. v_r in the direction AC by the law of vector addition.

A good familiar example in which a body has a velocity compounded of two velocities is to be obtained from the case of a wheel rolling along the ground. Any point on the wheel is moving around the axle and the axle is at the same time moving along parallel to the ground so that each point upon the wheel is actually describing a curved path—indicated in dotted lines in

Fig. 58. This curved path is called the *cycloid* and is also used by engineers in considering gear teeth. This curve can be drawn by rolling a half-crown along a ruler and resting a pencil against

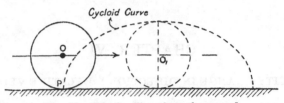

Fig. 58. Wheel rolling along the ground.

the edge of the coin. A milled coin such as the half-crown is better than a penny because it will not slip on the ruler. With a little practice a very smooth curve can be obtained.

Change of Velocity. Suppose that a point at A, Fig. 59, has a velocity v_1 at one instant and after a certain time it is at

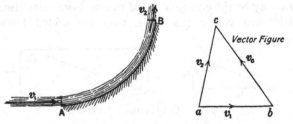

Fig. 59. Change of Velocity.

B and has a velocity v_2. Then the change of velocity v_c is defined as the velocity which would have to be compounded or combined with v_1 to give v_2. That is v_2 is the resultant of v_1 and v_c, or expressing this in vector notation we have

$$v_2 = v_1 \dplus v_c.$$

This problem arises in engineering calculations in considering the impact of water upon the vanes of a water wheel or turbine.

Numerical Example. *A jet of water moving with a velocity of* 80 *feet per second impinges upon a curved plate and has its direction turned through* 120°, *without altering its magnitude. What is the change in velocity?*

Referring to Fig. 60, we draw *ab* to a suitable scale to represent 80 feet per second and *ac* at 120° to it to represent 80 feet per second also and then join *bc*; then *bc* represents the change of velocity and if the diagram is drawn to scale *bc* will be found by

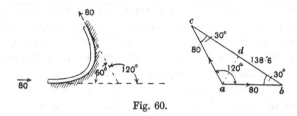

Fig. 60.

measurement to give about 138·6 feet per second. To find v_c by calculation without actually drawing the triangle to scale we draw *ad* perpendicular to *bc*; we then note that

$$cd = ac \cos 30° = 80 \times \cdot 866,$$

$\therefore bc = 2cd = 160 \times \cdot 866 = \underline{138\text{·}6 \text{ feet per second nearly.}}$

Relative Velocity. We now come to a very important portion of the subject which students often find rather difficult to understand and to which therefore we wish to give particular attention. In the ordinary way when we speak of velocities in a certain direction (say 4 miles an hour in a northerly direction) we leave out of consideration the fact that the earth is not fixed. Since the earth itself is rotating on its axis as well as moving through space at a very high velocity the actual velocity of any point is the combination of the velocity commonly referred to and that of the earth. We express this by saying that velocities as ordinarily measured are *relative* to the earth.

If we sit at the back of a dog-cart we can easily get the idea that the road is moving away from under us; that is because we regard ourselves as fixed and therefore relatively the road is moving away from us.

If, again, two trains are standing alongside in a railway station, and one starts moving, a person sitting in one train and looking at the other always has some doubts as to which of the trains is moving. Suppose that we are sitting in the train which we will call *A* and that the other is *B*. If *B* moves we have the sensation of moving in the opposite direction.

Now if two bodies A and B are both moving, *the velocity of B relative to A is the velocity which B would appear to have if A were regarded as stationary.*

Let us consider a fact that every observant reader will already have noticed, viz. that the rain splashes upon the window of a train or other moving vehicle are never vertical although the rain may be falling quite vertically; they are always inclined away from the direction of the train, i.e. they always seem to be coming towards the train as indicated in Fig. 61. In answer to the question as to why this is, we usually say that the relative velocity of the rain to the train is in that inclined direction; but that answer does not give very much

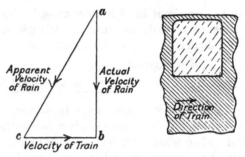

Fig. 61.   Relative Velocity.

enlightenment and the reader must realise for himself the meaning of it because mere mental assent to the assertion is useless. We will therefore make a very simple model as follows: on a piece of tracing paper draw a rectangle $ABCD$, Fig. 62, and draw a horizontal line XY upon a piece of ordinary drawing paper and take a number of points 1_T, 2_T, 3_T, 4_T, 5_T, etc. on this line at equal distances apart. Draw also a vertical line ZU and take upon it points 1_D, 2_D, 3_D, 4_D, 5_D also at equal distances apart. The points on the line XY represent successive positions of the lower right-hand corner of the train window and the points on the line ZU represent corresponding positions of the rain drop. Strictly, the rain drop may be moving with an acceleration so that the distances 1_D, 2_D; 2_D, 3_D, etc. may progressively increase in length, but the whole length ZU is so small that we may neglect this refinement. As a matter of fact the resistance to the movement of a rain drop makes its acceleration quite small.

Now place the rectangle $ABCD$, representing the window, on the tracing paper with the point C at the point 1_T and mark the point 1_D on ZU on the tracing paper; then shift C to 2_T and trace the point 2_R as shown in dotted lines on the tracing paper; then move to 3_T and trace the point 3_R and so on. The points on the tracing will then join up to an inclined line as shown dotted, and this is the direction of the relative velocity between the rain and train, i.e. the direction which the rain appears to have from the train. Consider any particular point say 4_R; by the time that

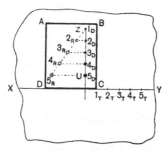

Fig. 62.   Relative Velocity.

the rain has fallen from 1_D to 4_D, 4_R has also moved to 4_D and the rain strikes 4_R in its new position; to get the apparent position of 4_R we set 4_D4_R *back* a distance equal to the distance which 4_R has moved in the given time.

Referring back to Fig. 61 if ab represents the actual velocity of the rain and cb represents the velocity of the train ac will represent the relative velocity of the train to the rain. This we may regard as a rule which we have proved experimentally; it will be found true for any numerical values which may be taken.

General rule for Relative Velocities.   With this preliminary explanation we will now give the general rule for relative velocities. Suppose that a point A, Fig. 63, is moving with a velocity v_A with reference to a certain plane in the direction indicated and that the point B is simultaneously moving with a velocity v_B with reference to the same plane in the direction indicated, A and B being positions of the points at the same instant.

To a convenient scale set out oa parallel to v_A to represent v_A in direction and magnitude and to the same scale set out ob to represent v_B in direction and magnitude and join ab, then *ab* is the velocity of *B* relative to *A*, i.e. *ab* is the velocity which *B* appears to have to a person moving with *A*; it is written r_{BA}. Similarly *ba* is the velocity of *A* relative to *B*.

It will be noted that this construction is different from that employed for finding the resultant of v_A and v_B; if the resultant had been required we should have drawn *ab'* to represent v_B as indicated in dotted lines and *ob'* would have given the resultant, and is the vector sum.

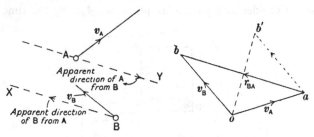

Fig. 63.   Relative Velocities.

It will be noticed that v_B is the vector sum of v_A and r_{BA}, i.e. using the vector notation

$$v_B = v_A + r_{BA}.$$

Therefore using ~ to indicate vector difference we have

$$r_{BA} = v_B \sim v_A.$$

Expressing this in words we see that *the velocity of B relative to A is the vector difference between the velocity of B and the velocity of A.*

The dotted lines *BX* and *AY* which are each parallel to *ab* are the paths which *B* and *A* appear to take from *A* and *B* respectively.

Numerical Examples.   (1) *If a train is running at 30 miles an hour, in what direction must a stone be thrown at a velocity of 60 feet per second to pass in through one open carriage window and out through the opposite window ?*

Referring to Fig. 64, the stone must have a velocity relative to the train in a direction *AB*, i.e. at right angles to the direction of motion of the train.

30 miles an hour = 44 feet per sec., so let *oa* represent 44 feet per second; draw *ab* at right angles to *oa* and with *o* as centre

draw an arc of radius representing 60 feet per second cutting
ab in *b*. This determines the point *b* and completes the triangle
of velocities. We want to find ∠*aob* to obtain the direction in
which the stone must be thrown.

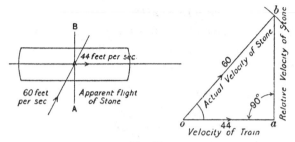

Fig. 64.

If we draw to scale we shall find that the ∠*aob* is about 43°;
by calculation we have

$$\cos aob = \frac{oa}{ob} = \frac{44}{60} = \cdot7333.$$

From tables we find *aob* = 42° 50'.

(2) *A ship A is steaming due N. at a speed of 10 miles an
hour ; when another boat B is due W. of A and at a distance
21 miles from it, B starts at a speed of 10 miles an hour in a N.E.
direction. What is the least distance apart that B will attain from
A and how long after starting will B be at its least distance from A ?*

This question is a little more difficult, but with the following
explanation the student should not have much difficulty in
following it.

Fig. 65 indicates the position of the boats at the first instant
under consideration.

We first draw the vector figure to obtain the relative velocity
of *B* to *A*. Draw *ob* in a N.E. direction to represent 10 miles
an hour and *oa* in a N. direction to represent 10 miles an hour
also. Then *ab* is the velocity of *B* relative to *A*, i.e. *ab* is the
velocity which *B* appears to have to a person on *A*. Now draw
BE parallel to *ab*; then if *A* were fixed *BE* would be the path
taken by the steamer *B*. The boat *B*, therefore, will appear from
A to move along the line *BE* with a velocity represented by *ab*.

If the △*oab* be drawn carefully to scale, *ab* will be found to
represent 7·65 miles an hour.

If we do not plot to scale we can calculate ab as follows:
Draw ox perpendicular to ab.

Then $$\frac{ax}{oa} = \sin 22\tfrac{1}{2}°,$$

i.e. $$ax = 10 \sin 22\tfrac{1}{2}°;$$
$$\therefore ab = 2ax = 20 \sin 22\tfrac{1}{2}°$$
$$= 7 \cdot 654 \text{ miles per hour.}$$

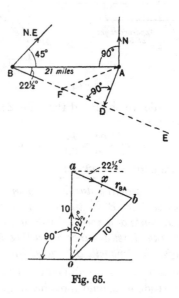

Fig. 65.

Suppose that after a given time, say one hour, B has arrived
at F in its apparent path BE; then BF will be 7·65 miles and
AF will be the distance of B from A at that instant; in other
words the distances from A to various points on BE give the
distances apart of the boats at various times.

The least distance apart of the boats will therefore be given
by AD, where AD is drawn perpendicular to BE. By measurement this should come to 8·04 miles.

By calculation we have

$$\frac{AD}{AB} = \sin 22\tfrac{1}{2};$$

$$\therefore AD = 21 \sin 22\tfrac{1}{2} = \underline{8 \cdot 036 \text{ miles.}}$$

Now the time taken for this will be the time for B to move to D at a speed of 7·65 miles an hour.

Now $BD = 19.4$ miles (by measurement),

(by calculation) $\dfrac{BD}{AB} = \cos 22\frac{1}{2}$;

$$\therefore BD = 21\cos 22\frac{1}{2} = 19.40 \text{ miles} ;$$

$$\therefore \text{Time} = \frac{19.40}{7.65} = 2.53 \text{ hours},$$

say 2·5 hours.

SUMMARY OF CHAPTER VI.

Velocities are combined by the law of vector addition.

The velocity of B relative to A is the velocity which B would appear to have if A were regarded as stationary; it is the vector difference between the velocity of B and the velocity of A.

EXERCISES. VI.

1. A railway train going at 30 miles an hour is struck by a stone moving horizontally at right angles to the train at a velocity of 33 feet per second. What are the magnitude and direction of the velocity with which the stone appears to meet the train?

2. A body is moving towards the north at 50 ft. per sec. In two secs. afterwards we find that it is moving towards the north-east at 60 ft. per sec. Find the magnitude of the added velocity.

3. A ship is sailing N.E. at 10 miles an hour, and to a passenger on board the wind appears to blow from the N. with a velocity of 14·14 miles. Find the actual velocity and direction of the wind.

4. Water enters a turbine wheel at an angle of 35° to the circumference, with a velocity of 80 ft. per sec. If the speed of the circumference of the wheel is 60 ft. per sec., find the velocity of the water relative to the wheel in magnitude and direction.

5. Two trains each 200 feet long are moving in parallel lines with velocities of 20 and 30 miles an hour in the same directions. How long will they be in passing?

6. Two trains pass one another moving in opposite directions on parallel lines of rail, with velocities of 45 and 60 miles per hour. The length of one is 420 ft. and of the other 350 ft. How long will they be in passing one another?

7. Two boats each 30 ft. long are rowed at 8 and 7 miles per hour respectively, the latter being 80 ft. ahead of the former. Find how long before it is bumped; also the time before the former draws level with it and the extra time necessary to pass it.

8. A train is travelling at a rate of 20 miles an hour and a man, sitting in a compartment with both windows open, observes a stone pass through both windows at right angles to the direction of the train. If the stone appears to move 20 feet per second to the man, with what velocity must it have been thrown?

9. A is travelling due N. at a constant speed. When B is due W. of A and at a distance of 21 miles from it, B starts travelling N.E. with the same constant speed as A. Determine graphically or otherwise the least distance which B will attain from A.

10. A cyclist is riding due W. at 12 miles an hour and the wind is blowing from the S.E. at $5\frac{1}{2}$ miles an hour. If the cyclist carries a small flag, in what direction will this flag fly? At what speed would the cyclist have to ride to make the flag fly due N.?

CHAPTER VII

KINETIC ENERGY AND MOMENTUM

Measurement of Kinetic Energy. We have already explained (p. 40) that kinetic energy is the amount of work stored in a body in virtue of its velocity but we have not yet explained how the kinetic energy can be measured.

Suppose that a body P, of weight W, starting from rest falls from A to B, Fig. 66, without overcoming any resistance. Then

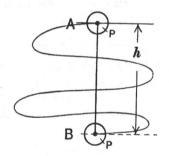

Fig. 66.

if h is the vertical distance moved, the weight W has done an amount of work upon the body equal to force × distance moved by the body in the direction of the force = Wh. Since no work has been spent in overcoming resistance, the whole of this work must be stored up in the body in the form of kinetic energy (K.E.), and since the body was originally at rest and possessed no kinetic energy it follows that its kinetic energy at the point B is equal to Wh,

i.e.
$$\text{K.E.} = Wh \quad \dots\dots\dots\dots\dots(1).$$

But we have already shown that for bodies falling freely under the action of gravity

$$v^2 = 2gh,$$ [formula p. 97]

i.e. $$h = \frac{v^2}{2g};$$

∴ we have $$\text{K.E.} = \frac{Wv^2}{2g} \dots\dots\dots\dots\dots(2).$$

This is a very important formula. In using it, we must note that it does not matter how the body has moved in obtaining this velocity; all that matters is that the body, somehow or other, has attained a velocity v. Then we say that its K.E. is $\frac{Wv^2}{2g}$.

We showed for example, on p. 37, that the work done by a force depends only on the straight distance in the direction of the force between the original or final positions of the body. If for instance the body had moved in the irregular path indicated in Fig. 66, the work stored in it would still have been Wh and therefore the kinetic energy would still be Wh, and since the kinetic energy depends only on the velocity, by definition it must be equal to $\frac{Wv^2}{2g}$ whatever be the path traversed.

The direction of the velocity will be different in the two cases, but that does not matter so far as kinetic energy is concerned.

Change in kinetic energy. Suppose that a body of weight W has at one instant a velocity u and at some subsequent instant it has a velocity v.

Then its kinetic energy has changed from $\frac{Wu^2}{2g}$ to $\frac{Wv^2}{2g}$.

∴ Change in K.E. $= \frac{W}{2g}(v^2 - u^2) \dots\dots\dots(3).$

Complete energy equation. We have shown on p. 40 that work cannot be destroyed and that the difference between the amounts of work done by the effort and the resistance must be equal to increase or decrease of the kinetic energy.

We therefore have

Work done by effort = Work done against resistance
+ Gain in kinetic energy,

i.e. $$E_E = E_R + \text{K.E. gained} \dots\dots\dots\dots(4).$$

One of the best examples in practice of the use of kinetic energy arises in the use of flywheels to steady the motion of machines. We shall deal with flywheels later under rotating bodies.

Numerical Examples on Kinetic Energy. (1) *A bullet weighing* 3 *ounces is discharged from a rifle with a velocity of* 1200 *feet per second. How much kinetic energy does it possess and how far will it be able to move a body the resistance to whose motion is* 10 *tons if we neglect the energy lost in the impact?*

The weight of the body in lbs. $= W = \dfrac{3}{16}$;

$$\therefore \text{K.E.} = \frac{Wv^2}{2g} = \frac{3}{16} \times \frac{1200 \times 1200}{32 \cdot 2}$$
$$= 83{,}800 \text{ ft.-lbs. nearly.}$$

If the resistance to the motion of a body is R lbs., the work done in moving the body a distance s feet in opposition to the resistance is equal to Rs ft.-lbs.

In our present case $R = 10$ tons $= 22{,}400$ lbs.

$$\therefore 22{,}400s = 83{,}800,$$
$$s = \frac{83{,}800}{22{,}400} = 3 \cdot 74 \text{ feet.}$$

We wish to warn the student that the above calculation is chiefly of academic interest and of value as an exercise in applying the formulae. As a matter of fact considerable energy is absorbed in the impact, being converted into the thermal form of energy, and the resisting force will not be constant.

(2) *A train weighing* 100 *tons gets up a speed of* 30 *miles an hour in* 1 *mile from rest on the level, the air and other resistances being equivalent to a force of* ¾ *ton. What constant tractive effort is required?*

In this case original K.E. $= 0$.

After one mile $\qquad \text{K.E.} = \dfrac{Wv^2}{2g}$,

$v = 30$ miles an hour $= 44$ feet per sec.,

$$\therefore \text{K.E.} = \frac{100 \times 2240 \times 44 \times 44}{2 \times 32 \cdot 2}$$
$$= 6{,}734{,}000 \text{ ft.-lbs. nearly.}$$

The distance travelled in getting up a speed of 44 feet per second is 1 mile, i.e. 5280 feet.

Therefore if the effort is F lbs. we have

Work done by effort = Work against resistance + Gain in K.E.

$$\text{Resistance} = \frac{3}{4} \text{ ton} = \frac{3 \times 2240}{4} \text{ lbs.} = 1680 \text{ lbs.}$$

$$\therefore F \times 5280 = 1680 \times 5280 + 6,734,000$$

or $\qquad (F - 1680)\, 5280 = 6,734,000,$

$$F - 1680 = \frac{6,734,000}{5280}$$

$$= 1275 \text{ lbs.}$$

$$\therefore F = 1275 + 1680 = \underline{2955 \text{ lbs.}}$$

(3) *Taking the numerical example worked on p. 48, find the maximum velocity and the velocity at the end if the initial velocity was 10 feet per second and the body weighs 1000 lbs.*

We have given that the gain in K.E. up to the point of maximum velocity = 2450 ft.-lbs.

$$\text{Initial K.E.} = \frac{Wu^2}{2g} = \frac{1000 \times 10 \times 10}{2 \times 32 \cdot 2} = 1553 \text{ ft.-lbs.};$$

$\therefore$ K.E. at maximum velocity = $1553 + 2450$ ft.-lbs. = 4003;

$\therefore$ if v is the maximum velocity

$$\frac{Wv^2}{2g} = 4003,$$

i.e. $\qquad \dfrac{1000v^2}{2 \times 32 \cdot 2} = 4003,$

$$v^2 = \frac{2 \times 32 \cdot 2 \times 4003}{1000} = 258 \text{ nearly};$$

$$\therefore v = \sqrt{258} = \underline{16 \text{ ft. per sec.}}$$

We next find the velocity at the end of the motion as follows.

At the end, the excess work which appears as kinetic energy was found to be 1000 ft.-lbs.

Therefore K.E. at end = K.E. at beginning + work added

$$= 1553 + 1000$$

$$= 2553 \text{ ft.-lbs.}$$

$$\therefore \frac{Wv^2}{2g} = 2553,$$

$$\therefore v^2 = \frac{2g \times 2553}{1000} = \frac{2 \times 32 \cdot 2 \times 2553}{1000}$$

$$= 164 \text{ nearly};$$

$$\therefore v = \sqrt{164} = \underline{12 \cdot 8 \text{ feet per sec.}}$$

The connection between Force and Acceleration. Suppose that a force F is acting upon a body weight W. It has been found experimentally that the body will be given a uniform acceleration a in the direction of the force and that a bears the same ratio to g the gravity acceleration as F bears to W.

We therefore have the rule

$$\frac{a}{g} = \frac{F}{W} \quad \dots\dots\dots\dots\dots (5),$$

or

$$F = \frac{Wa}{g} \quad \dots\dots\dots\dots\dots (6).$$

This law is one of the most important in the whole range of mechanics and the student must master it before he can hope to appreciate the interest and importance of the subject. It was discovered by **Newton** and is often called the second law of motion although it is usually expressed in different language. We have already considered Newton's other two laws of motion and will summarise them again a little later. For the present we will endeavour to become familiar with these formulae.

Suppose that a body of weight W is moved by a force F a very short distance s in the direction of F and that its velocity at the beginning of the distance is u and at the end is v.

Then work done $= F \cdot s$.

If this all goes in increasing the K.E. we have

$$F \cdot s = \text{gain in K.E.}$$

$$= \frac{W}{2g} (v^2 - u^2) \quad \dots\dots\dots\dots (7).$$

Now if s is so short that the force F is constant over it and that the acceleration is also constant, we have by formula (7), p. 96,

$$v^2 = u^2 + 2as,$$

i.e.

$$v^2 - u^2 = 2as.$$

Putting this in (7)

$$F \cdot s = \frac{W}{2g} \cdot 2as,$$

i.e.

$$\therefore F = \frac{Wa}{g} \quad \dots\dots\dots\dots\dots (8).$$

We can thus deduce the result from the principle of the conservation of energy.

Numerical Examples. (1) *The piston of an engine weighing 20 lbs. is given a retardation of 6 feet per second per second. What backward pressure will be acting on the piston?*

We shall discuss the crank and connecting-rod mechanism of an ordinary engine in a later chapter (p. 258) and those students who would like to understand this question with special reference to its application are recommended to refer to that description.

Putting our values in

$$P = \frac{Wa}{g} = -\frac{20 \times 6}{32 \cdot 2} \quad \begin{array}{l}(-\text{ indicates back-}\\ \text{ward pressure)}\end{array}$$

$$= 3 \cdot 73 \text{ lbs.} \quad Ans.$$

(2) *A train weighing 100 tons gets up a speed of 30 miles an hour in 1 mile from rest on the level, the air and other resistances being equivalent to a force of ¾ ton. What constant tractive effort is required?*

We have already worked this example on p. 115 from the work point of view; now let us work it from that of the acceleration.

We have $\qquad v^2 = u^2 + 2as,$

$v = 44$ feet per sec., $u = 0$, $s = 5280$ feet,

$$\therefore 44 \times 44 = 10560a,$$

$$a = \frac{44 \times 44}{10560} = \cdot 183 \text{ feet per sec.};$$

$$\therefore \text{Resultant force} = \frac{Wa}{g} = \frac{100 \times 2240 \times \cdot 183}{32 \cdot 2}$$

$$= 1274 \text{ lbs.}$$

Now resultant force = effort − resistance,

i.e. $\qquad\qquad 1274 = F - 1680;$

$$\therefore F = \text{tractive effort} = \underline{2954 \text{ lbs.}} \quad Ans.$$

We think that as a general rule the student will find the work method of solving problems easier to deal with than the acceleration method, but it is somewhat a matter of individual taste.

(3) *Take example (2) and find the effort required to give the same speed in the same distance up an incline of 1 in 100.*

Solution (i). By acceleration.

In this case there is in addition to the air and like resistances a resistance equal to the resolved component of the weight of the train in the direction of the tractive effort.

This is an example of the inclined plane and we proved (p. 58) that adb, Fig. 67, is the triangle of forces.

$$\therefore \text{Component down plane} = \frac{100 \times ab}{ad} = 1 \text{ ton nearly}$$
$$= 2240 \text{ lbs.}$$

$\therefore$ Resistance now $= 2240 + 1680 = 3920$ lbs.

Resultant force up plane $= 1275$ lbs. (as before).

$$\therefore \text{Tractive effort} = 3920 + 1275$$
$$= \underline{5195 \text{ lbs.}}$$

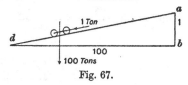

Fig. 67.

Solution (ii). By work equation.

When the train goes 1 mile its weight is lifted by an amount very nearly equal to $\frac{1}{100}$ mile $= 52 \cdot 8$ feet.

$\therefore$ Work done $= 100 \times 2240 \times 52 \cdot 8$ ft.-lbs. against gravity
$$= 5280 \times 2240;$$

$\therefore$ Work done by effort
$$= \text{Work done against resistance} + \text{gain in K.E.},$$

i.e. $F \times 5280 = 1680 \times 5280 + 2240 \times 5280 + 6{,}734{,}000,$

$\therefore 5280 \{F - (1680 + 2240)\} = 6{,}734{,}000,$

$$F - 3920 = \frac{6{,}734{,}000}{5280} = 1275 \text{ lbs.},$$

i.e. $F = 1275 + 3920 = \underline{5195 \text{ lbs.}}$

Momentum. We have seen that when a body is moving with a certain velocity, it has a certain amount of kinetic energy. It is said also to possess *momentum*, the amount of momentum being defined as follows. If a body of weight W lbs. is moving with a velocity v feet per second, its momentum is equal to $\dfrac{Wv}{g}$, i.e. $\dfrac{Wv}{32 \cdot 2}$ in lb.-ft.-second units.

Momentum is a vector quantity because it has direction as well as magnitude.

We can therefore combine momenta by the law of vector addition. Change of momentum is found in exactly the same manner as change of velocity (p. 104).

Numerical Example. *What is the momentum possessed by the bullet referred to in example* (1) *on p.* 115 ?

$$\text{Wt.} = 3\,\text{oz.} = \frac{3}{16}\,\text{lbs.,}\quad v = 1200\,\text{feet per second};$$

$$\therefore \text{Momentum} = \frac{3 \times 1200}{16 \times 32\cdot2} = 7\,\text{units.}$$

Dimensional equations. We can find the dimensions of the units in any formula by writing a dimensional equation as follows:

$$\text{Momentum} = \frac{Wv}{g} = \frac{\text{lbs.} \times \dfrac{\text{ft.}}{\text{sec.}}}{\dfrac{\text{ft.}}{\text{sec.}^2}}$$

$$= \frac{\text{lbs.} \times \text{sec.}^2 \times \text{ft.}}{\text{ft.} \times \text{sec.}}$$

$$= \text{lbs.} \times \text{seconds.}$$

In these dimensional equations we cancel out dimensions according to ordinary rules of fractions, and students should write such an equation whenever they are not certain as to whether or not their formula is in the right dimensions. Mere numerical coefficients are not counted.

As another example take kinetic energy:

$$\text{K.E.} = \frac{Wv^2}{2g} = \frac{\text{lbs.} \times \left(\dfrac{\text{ft.}}{\text{sec.}}\right)^2}{\dfrac{\text{ft.}}{\text{sec.}^2}}$$

$$= \frac{\text{lbs.} \times \text{ft.}^2 \times \text{sec.}^2}{\text{sec.}^2 \times \text{ft.}}$$

$$= \text{lbs.} \times \text{ft.}$$

We know that energy should be in work units and the fact that this comes to lbs. × ft., i.e. in work units, shows us that the formula is of the right order.

The importance of Acceleration in Traction Problems. In all branches of traction engineering—railways, tramways and motor cars—the question of acceleration is of the greatest possible importance and students who wish to specialise later in any of these branches should make themselves quite clear on this subject of the connection between effort or force and acceleration.

Electric traction is supplanting steam traction for suburban traffic, not because it is less expensive but because the electric trains can get up speed more quickly or in other words they can have a greater acceleration. If we have a fixed resultant effort F available, the acceleration a is given in terms of the weight W to be moved by the relation

$$a = \frac{Fg}{W},$$

and the smaller we make the weight W, the larger the acceleration a will become. This is why we want to keep down the weight as much as possible if we want to start quickly; at the same time the weight must be enough to get sufficient grip upon the rail or road to prevent the driving wheels from slipping; this is allowed for in electric trains by having motors on several of the carriages. If we have two vehicles such as bicycles exactly similar but very different in weight, say one made of aluminium and one of steel, we speak of the heavier one as more difficult to push even on the level; we mean really that it is more difficult to accelerate or start. There is practically no difference in the force required to keep the heavy and the light one moving at a given speed once they have been started. Modern traction is principally concerned with the question of getting up speed—and, as far as brake problems go, of slowing down—and to deal with these problems we must know how to calculate the acceleration when we know the resultant effort and the weight of the body.

SUMMARY OF CHAPTER VII.

Kinetic Energy (K.E.) $= \dfrac{Wv^2}{2g}$.

Change in K.E. $= \dfrac{W}{2g}(v^2 - u^2)$.

Work done by effort = Work done against resistance + Gain in kinetic energy.

$$F = \frac{Wa}{g}.$$

Momentum $= \dfrac{Wv}{g}$.

EXERCISES. VII.

1. A body of weight 10 lbs. moves with a linear velocity of 800 ft. per min. Find its kinetic energy in ft.-lbs.

2. A train weighing 150 tons is running on a level road at 30 miles per hour. The resistances are equal to 12 lbs. per ton. If steam be shut off how far will the train run before coming to rest? Give answer in yards.

3. A body weighing 100 lbs. increases its velocity from 25 to 35 yds. per sec. Find the increase in its kinetic energy.

4. In a Fly Press the radius at which the balls revolve (there are two each weighing 12 lbs.) is 10 ins. and the number of revolutions per min. is 150. If the die be brought to rest after stamping through a piece of metal ⅛ in. thick, what is the average force resisting the blow?

5. A shot weighing 6 lbs. leaves the mouth of a gun with a velocity of 1000 ft. per sec.; determine the number of ft.-lbs. of energy accumulated in it and the mean pressure exerted by the exploded powder behind it if the length of the bore is 5 ft.

6. A body weighing 108 lbs. is placed on a smooth horizontal plane, and under the action of a certain force describes from rest a distance of 11⅛ ft. in 5 secs. Find the force in lbs.

7. The rim of a flywheel weighs 9 tons and its mean linear velocity is 40 ft. per sec.; how many ft.-tons of work are stored up in it? If it is required to store the additional work of 9 ft.-tons what should be the increase in velocity?

8. A train weighing 50 tons is impelled along a horizontal road by a constant force of 550 lbs.; the frictional resistance is 8 lbs. per ton; what velocity will it have after moving from rest for 10 mins., and what distance will it describe in that time?

9. A car weighing 2½ tons and carrying 40 passengers of average weight 145 lbs. each is travelling on a level rail at 6 miles per hour. What is the momentum?

10. In the previous example what average force must be exerted to bring the car to rest in 2 seconds, and if that force is constant what distance will the car travel before it comes to rest?

11. A ship weighing 2500 tons is propelled at 20 knots (1 knot = 6080 ft. per hour) by engines of 8000 H.P. Estimate the distance which will be traversed by the ship whilst an amount of energy is developed by the engines equal to the kinetic energy of the ship.

12. A weight of 50 lbs. is moving at a speed of 15 feet per second and it is acted upon for 20 seconds by a force of 20 lbs. in the direction of motion. What is the distance moved through during the time?

13. A train weighing 250 tons is moving at 40 miles per hour and is stopped in 10 seconds. What is the average force causing stoppage?

14. A planing machine table weighs 2 tons and has a retardation at the end of its stroke of 3 feet per second per second. What thrust will this cause on the driving mechanism?

15. When starting, a locomotive exerts a tractive force of 4 tons upon a train weighing 200 tons. Calculate the acceleration (neglecting friction), and the velocity after 1 minute.

16. A piston and rod and cross-head weigh 330 lbs. At a certain instant, when the resultant total force due to steam pressure is 3 tons, the piston has an acceleration of 370 feet per second per second in the same direction. What is the actual force acting on the cross-head?

CHAPTER VIII

NEWTON'S LAWS OF MOTION: IMPACT

WE will now consider collectively Newton's Laws of motion which are the foundation of the whole scientific treatment of mechanics.

They may be enunciated as follows:

1. *A body continues in a state of rest or uniform motion in a straight line unless it be acted upon by some external force.*

2. *The rate of change of momentum is proportional to the force applied and takes place in the direction of the force.*

3. *To every action there is an equal and opposite reaction.*

1. The first law is sometimes called the law of *inertia*; inertia being the property of a body which resists a change in its state of rest or motion. It follows from this law that if there is a resultant force acting upon a body, it must either change its velocity if it is already moving or else start moving if it is stationary; in either case the body will be given an acceleration, which may be negative, i.e. a retardation. In all engineering problems dealing with bodies which from their very nature must be stationary—e.g. structures such as bridges, roofs, dams, etc.—we know from this law that all the forces acting must neutralise each other, or in the language of mechanics their resultant must be zero.

This law cannot be rigorously demonstrated experimentally because it is impossible for us to move bodies without external forces being brought into play; we have referred to these as passive resistances. A stone thrown along a road soon comes to rest on account of the forces—called frictional forces—caused by the roughness of the road, but if thrown along a surface of ice which has very little friction the stone will run for a very long way. These frictional forces are the bugbear of the engineer; he has for centuries been trying to make them as small as possible but he can never get rid of them altogether. If he could, the world would be a very different place. As a matter of fact frictional resistances, which we will deal with in detail later, are of great value in certain cases. In frosty weather for instance we throw sand down upon the roads to increase the friction because without it we should not be able to get sufficient grip to propel our vehicles. What we should like to be able to do is to bring frictional forces into play when they are useful and eradicate them when they are not, but natural phenomena will not change for our convenience; all that we can do is to study them as closely as possible in order to use them to our greatest possible advantage.

2. This is Newton's way of expressing the law that we have reduced to symbols in the form $F = \dfrac{Wa}{g}$ for the case in which the weight does not change. We will consider that case in detail. Suppose that in a very short time t seconds the velocity of a body of weight W changes from v feet per second to v' feet per second. Its change of momentum is equal to $\dfrac{Wv}{g} - \dfrac{Wv_1}{g} = \dfrac{W}{g}(v - v_1)$ and this takes place in a time t.

∴ Change of momentum per second

= Rate of change of momentum

$$= \frac{W}{g}(v - v_1) \div t = \frac{W}{g} \cdot \frac{(v - v_1)}{t}.$$

Now $\left(\dfrac{v - v_1}{t}\right)$ is the rate of change of velocity and this we have called the acceleration (a), so that we have

Rate of change of momentum $= \dfrac{Wa}{g}$.

Newton's law states that the rate of change of momentum is *proportional* to the force applied; whereas in our formula we make it *equal* to the force applied. This is because we choose our units of force and momentum so that the proportionality becomes an equality.

3. This law is very important and will become more clear if we give some explanatory considerations.

Take the case of a weight hung on the end of a rod, Fig. 68; the result of that action will be to cause the rod to stretch. The amount of stretching will be very small but it can be measured by delicate instruments called "extensometers." This stretching brings into play forces between the molecules of the rod tending to resist the motion. These molecular forces are called *stresses* and act across every section that we consider.

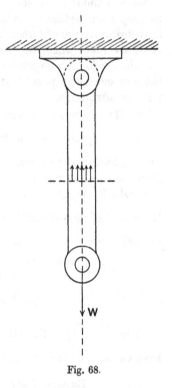

The stresses increase with the amount of the stretching, which will continue until the resultant of the stresses is equal to the weight W; this resultant of the stresses is the reaction which is equal and opposite to the action, i.e. the weight. If the reaction contributed by the stresses is not equal to the weight, there will be a resultant force acting upon the rod below the section under consideration. From the first law this must cause the change of state of rest of the rod, i.e. must start it moving. This is exactly what happens when the load is so great that the rod breaks. For every material there is a certain maximum stress that it is capable of calling into play so that the resultant stress can never be more than a certain amount; if therefore the weight is greater than this, motion must take place and the rod fractures.

Fig. 68.

When the load is removed the rod returns to its original length (unless the material is not perfectly elastic*), the return movement showing the existence of the stresses.

As another example take the case of a man striking his fist against a wall. The wall presses just as hard on the man's hand

* See p. 139.

as his hand presses on the wall and the feeling of pain which the man experiences is a proof of the existence of the reaction.

Next take the case of a traction engine pulling along a truck. The truck pulls back on the engine just as much as the engine pulls on the truck. How then is it that the truck goes along at all? We may answer that the truck does not move relatively to the engine and that if there were not equality between the forces between them there would be a resultant which would cause relative motion.

At the same time it is true that the engine must exert a greater effort than the force with which it pulls the truck. Referring to Fig. 69 let F be the effort which the engine exerts

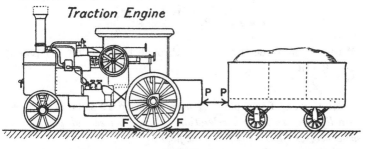

Fig. 69.

upon the ground; it is only by means of the grip upon the ground that the engine can pull; that is why the driving or back wheels are usually roughened. A traction engine would be absolutely useless upon ice because the wheels would merely slip round. If you watch the locomotive of a heavy train start you will usually notice that the driving wheels will slip and buzz round; the driver then operates a device for projecting sand under the driving wheels to increase the grip. The ground has to be able to exert the same effort F as a reaction or else, as we have seen, slip occurs.

Now part of this force will be spent in overcoming the resistances in the engine and accelerating it and part in overcoming the resistances in the truck and accelerating it. The force P therefore which has to be transmitted across the coupling is less than the force F which the engine has to exert upon the ground. One form of reply therefore to the question as to how it is that the engine can pull the truck is that the engine only exerts the same force upon the truck as the truck exerts upon the engine but that the engine exerts upon the ground a greater force than the resistance to the motion of the truck.

Let us consider this problem in somewhat greater detail.

Let R_E and R_T be the resistances to motion of the engine and truck respectively and let W_E and W_T be their weights. These resistances will depend upon the weights and the velocity to some extent and can of course only be found accurately by experimental determination for any given vehicle.

We then have

Total effort = Effort spent in overcoming resistance

+ Effort used in accelerating,

i.e. $$F = R_E + R_T + \frac{W_E \cdot a}{g} + \frac{W_T \cdot a}{g}.$$

Of this P the force transmitted through the coupling is equal to $R_T + \dfrac{W_T \cdot a}{g}$.

Numerical Example. *A man weighing 12 stone is going up in a lift which has an acceleration of 3 feet per second per second; what pressure does he exert on the floor?*

In this case the floor of the lift is exerting sufficient force to lift the man and in addition give him an acceleration of 3 ft. per sec.^{2}

$$\text{Force required for acceleration} = \frac{Wa}{g}$$

$$= \frac{12 \times 14 \times 3}{32 \cdot 2} \text{ lbs.}$$

$$= 15 \cdot 6 \text{ lbs. nearly.}$$

Therefore total upward pressure $= 12 \times 14 + 15 \cdot 6$

$$= \underline{183 \cdot 6 \text{ lbs.}}$$

Since the pressure of the floor upon the man must be exactly equal to his pressure on the floor, the man must exert a pressure of 183·6 lbs.

As the lift slows down, the acceleration is negative so that the pressure is less than the man's weight. The same occurs when the lift starts downwards and accounts for the unpleasant feeling which often accompanies a quick-stopping lift.

Impact and Impulse. Up to the present we have considered only the cases in which the forces acting are gradually applied and act over a considerable length of time. In some cases, however, the forces act over an extremely short time, as in an explosion or the blow of a hammer, and then we require to be able to estimate the force of the blow. Such suddenly applied forces are called *impulsive forces* and it is very difficult to calculate accurately the maximum force produced by a blow. All that we can do is to find the average value of the force if we know the short time during which the force or impulse acts. It is very necessary to realise the difference between the average force and the maximum. The determination of the maximum force of an impact is a very troublesome and advanced problem, but a knowledge of the average force is of considerable value to us in some problems.

Suppose that a body of weight W lbs. moving with a velocity v feet per second is suddenly brought to rest in t seconds. The average force F produced by the blow will be equal to the rate of change of momentum, i.e. to the momentum destroyed per second. But the original momentum was $\frac{Wv}{g}$ and it was destroyed in t seconds, so that $\frac{Wv}{gt}$ is the momentum destroyed per second.

Therefore we have average force produced

$$= F = \frac{Wv}{gt} \text{ lbs.} \quad \ldots\ldots\ldots\ldots\ldots (1).$$

Numerical Example. *A hammer weighing 2 lbs. and having a velocity of 30 feet per second strikes a blow lasting $\frac{1}{100}$ second. What force is produced by the blow?*

Substituting directly in formula (1) we have

$$F = \frac{2 \times 30}{32 \cdot 2 \times \frac{1}{100}} = \frac{6000}{32 \cdot 2} = \underline{186 \text{ lbs.}}$$

This question of impact is of very great importance and through loose use of language many people have wrong ideas about it. For instance a man may ask what force he can exert with a certain hammer; the correct answer is that you cannot tell because you must know the time that the blow lasts before you can know the force of the impact. If the hammer loses its momentum very quickly it strikes a heavy blow, but if it takes comparatively long to do so only a light blow results. A certain hammer will strike a very much heavier blow upon cast iron than it will upon wood and it will strike a much harder blow upon wood than upon sand. A carpenter does not use a heavy hammer for a chisel because wood is soft and does not require a very heavy force to cut it, the tool therefore has a wooden handle to absorb some of the shock and make the cut longer; but when tooling a very hard material a greater force is required so that as "snappy" a blow as possible is delivered. When we wish to avoid percussive forces we provide some device such as springs for making the force act over a longer time; for this reason we hang our carriages upon springs to deaden the effect of impulsive forces.

Equality of Momentum before and after impact. If two bodies collide or an explosion occurs between them, there are forces acting between them and by Newton's third law the forces are equal and opposite. Therefore the rates of change of momentum of the two bodies are equal and opposite. Thus one body gains in momentum in any direction as much as the other loses or the total momentum of the two bodies in any direction remains unchanged. This is sometimes called the law of the *conservation of momentum* and is an extremely important principle. Momentum cannot be destroyed by impact; its effects can only be got rid of by swamping it, i.e. by transferring it to a body whose weight is so great that the resulting change in velocity is negligible. If for instance a stone hits the ground, it loses its own velocity and therefore its momentum but it communicates the momentum to the earth. The weight of the earth is, however, so immense that the resulting velocity is so small that for all practical purposes it may be taken as nothing. We have constant examples of the reactive forces caused by a sudden change of momentum; we will consider some of them in detail.

Impact of bodies. Referring to Fig. 70, let one body of weight W possessing a velocity u collide with another body of weight W_1 moving in the same direction with a velocity u_1. Then the total momentum before impact

$$= m = \frac{Wu}{g} + \frac{W_1 u_1}{g} \quad \ldots\ldots\ldots\ldots\ldots (2).$$

If the velocities after impact are respectively v and v_1 in the same direction we shall have that after impact

$$m = \frac{Wv}{g} + \frac{W_1 v_1}{g} \dots \dots \dots \dots (3).$$

Since these must be the same we have, cancelling out g which is common throughout,

$$Wu + W_1 u_1 = Wv + W_1 v_1 \dots \dots \dots (4).$$

This equation alone is not sufficient to determine the velocities after impact unless the bodies are "inelastic," i.e. they do not rebound. The accurate treatment of the impact of elastic bodies such as billiard balls is very difficult and beyond the scope of the present book.

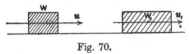

Fig. 70.

Restricting therefore our consideration to bodies which do not rebound we shall have the two bodies going on together at the same velocity after impact; if this common velocity is v we shall have in equation (4)

$$Wu + W_1 u_1 = Wv + W_1 v$$
$$= (W + W_1)\, v,$$
$$\therefore v = \left(\frac{Wu + W_1 u_1}{W + W_1} \right) \dots \dots \dots \dots (5).$$

Loss of energy at impact. Although there is no loss of momentum at impact there will always be a loss of kinetic energy. This loss of energy is evidenced by the noise produced by the impact and also by the heat produced; bullets for instance become very hot on impinging against anything.

The loss of energy can be expressed in formulae as follows: taking as before the case in which there is no rebound, we have

$$\text{Total Kinetic Energy before impact} = \frac{Wu^2}{2g} + \frac{W_1 u_1{}^2}{2g} \quad ..(6),$$

$$\text{Total Kinetic Energy after impact} = \frac{(W + W_1)\, v^2}{2g} \quad \dots \dots (7)$$

$$= \text{(from (5))} \; \frac{(W + W_1)}{2g} \left(\frac{Wu + W_1 u_1}{W + W_1} \right)^2$$

$$= \frac{(Wu + W_1 u_1)^2}{2g\,(W + W_1)} \quad \dots \dots \dots \dots \dots \dots \dots (8).$$

9—2

$\therefore$ subtracting (8) from (6) and multiplying out, we have

Loss of K.E.

$$= \frac{Wu^2}{2g} + \frac{W_1 u_1^2}{2g} - \frac{W^2 u^2 + 2WW_1 uu_1 + W_1^2 u_1^2}{2g(W + W_1)}$$

$$= \frac{1}{2g}\left\{\frac{Wu^2(W+W_1)+W_1 u_1^2(W+W_1)-(W^2 u^2 + 2WW_1 uu_1 + W_1^2 u_1^2)}{(W+W_1)}\right\}$$

$$= \frac{1}{2g}\left\{\frac{WW_1 u^2 + WW_1 u_1^2 - 2WW_1 uu_1}{W+W_1}\right\}$$

$$= \frac{WW_1}{2g(W+W_1)}\{u^2 - 2uu_1 - u_1^2\}$$

$$= \frac{WW_1}{2g(W+W_1)}(u - u_1)^2 \quad \dots\dots\dots\dots\dots\dots\dots(9).$$

Now $(u - u_1)$ is the relative velocity between the two bodies so that we have

$$\text{Loss of K.E.} = \frac{WW_1}{2g(W+W_1)} \times \text{square of relative velocity.}$$

It should be noted that if one of the bodies is moving in a direction opposite to that of the other, one of the velocities should be considered negative.

This principle is useful in questions dealing with the waste of energy due to a sudden contraction in a water pipe.

Numerical Example. *Water is flowing through a pipe and has a velocity of 3 feet per second until it meets a sudden contraction when the velocity is suddenly increased to 6 feet per second. How much kinetic energy is lost per pound of water flowing?*

In this case the change of velocity is not absolutely instantaneous in practice because the water will curl round somewhat as shown in Fig. 71, but experiments have shown that the present method of treatment based upon impact formulae gives results that are approximately correct.

Now the velocity increases and therefore the kinetic energy increases; the pressure, however, of the water will diminish. If the change of section were gradual this diminution of pressure would be such as to keep the total energy constant, but with the abrupt change there will be loss of energy and the pressure will therefore be less still.

In our example $W = 1$ lb. and W_1 is very large, because the pipe must be fixed to something very heavy else it would be pushed along.

$$\therefore \text{ Energy lost per pound} = \frac{1 \times W_1}{2g \, (1 + W_1)} \, (3 - 6)^2.$$

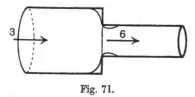

Fig. 71.

Now $\dfrac{W_1}{1 + W_1} = 1$ to all practical purposes if W_1 is very great; if for instance $W_1 = 10,000$ lbs.,

$$\frac{W_1}{1 + W_1} = \frac{10,000}{10,001};$$

$$\therefore \text{ Energy lost} = \frac{1}{2 \times 32 \cdot 2} \times (-3)^2$$

$$= \frac{9}{64 \cdot 4} = \underline{\cdot 14 \text{ ft.-lbs.}}$$

Recoil of guns, etc. Everyone who has used a rifle knows that it recoils or kicks back as the shot is fired. This is because before the explosion the shot had no velocity and therefore no momentum; it is suddenly given a velocity and momentum at the explosion, and since the total momentum before and after the impulse must be zero, the rifle is given an equal and opposite momentum which will drive the rifle backwards to an appreciable extent if it is not securely held.

An exactly similar effect is noticeable in the case of hose pipes. Firemen have to hold the hose pipe quite firmly, otherwise the pipe would jump backwards. This recoil was made use of in the simplest and earliest form of steam-engine (Hero's engine) and in an early form of water wheel known as "Barker's Mill," the modern form of which is the sprinkler used to water lawns. In this form the water rushes out and drives the radiating arms backwards.

Experiment. A very simple and instructive form of Hero's engine can be made as follows: Take a piece of glass tubing such as is used very largely in chemical apparatus and by heating in a Bunsen burner and drawing down obtain a piece about ½ inch long with a thin neck at each end, as shown in Fig. 72 (a). Now cut off one end as indicated in dotted lines and close this end by heating in the flame and then blow a bulb at the end as shown at (b). Next soften the glass just above the bulb at A and bend the stem over to resemble a glass retort and then bend over at right angles to the plane of the paper at B, about half an inch from the top. Our reaction steam turbine is now complete.

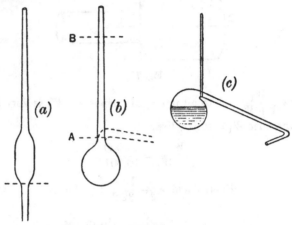

Fig. 72.

We have next to get some water into it. To do this warm it carefully without making it very hot and thus drive out some of the air from it; then dip the open end into water and the bulb will fill partly as the air cools. Now hang it up as indicated in (c) by means of a piece of thread, or better by means of a stump of wire joined to a piece of thread, and hold it over a gas flame, taking care not to burn the thread. The water will then boil and the steam rushes out at the end and makes the "engine" buzz round in merry fashion.

This question of recoil is of very great importance in the case of large guns, particularly those mounted on ships. In the design of battleships great care has to be taken that the stability is sufficient to bear the tremendous backward thrust of a broadside.

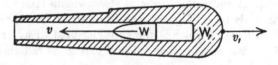

Fig. 73.

Referring to Fig. 73, let W be the weight of the shell and W_1 be the weight of the gun and let v be the velocity with which the shell is driven forward as a result of the explosion; then the gun will be driven backward with a velocity v_1. Then if we neglect the momentum of the gases resulting from the explosion, the backward momentum of the gun must be equal to the forward momentum of the shell,

i.e. $$\frac{W_1 v_1}{g} = \frac{Wv}{g},$$

or $$v_1 = \frac{Wv}{W_1}.$$

Numerical Example. *A gun weighing 40 tons fires a shell weighing 100 lbs. with a velocity of 1500 feet per second. What is the velocity of recoil ?*

In this case $W_1 = 40$ tons $= 40 \times 2240$ lbs.,

$W = 100$ lbs.,

$v = 1500$ feet per second;

$$\therefore v_1 = \frac{Wv}{W_1} = \frac{100 \times 1500}{40 \times 2240} \text{ feet per second}$$

$$= 1\text{·}67 \text{ feet per second.}$$

The action of a pile-driver. The operation of driving a pile into mud or soft earth gives us a familiar engineering application of the principles of impact. The pile A, Fig. 74, of wood, or nowadays of reinforced concrete, is driven into the mud by blows with a hammer "tup" or "monkey" B. This tup has in one form of pile-driver a hook C which is weighted so as normally to engage an endless chain D which moves upwards and carries the tup with it. The hook then meets a releasing or "trip" device E which is suspended by a rope so as to be adjustable in height; the tup is then freed from the chain and drops on to the pile, thus driving it in to an extent dependent upon the resistance of the mud or soil.

Let W be the weight of the pile and W_1 that of the tup and let h be the height through which the latter falls.

Then its velocity v_1 is given by $v_1{}^2 = 2gh$,

or $$v_1 = \sqrt{2gh};$$

therefore if the tup does not rebound we shall have that if v is the velocity with which the pile and tup move,

Momentum after impact $= (W + W_1)\, v$,

Momentum before impact $= W_1 v_1$;

$$\therefore v = \frac{W_1}{W + W_1} \times v_1.$$

If the resistance to the pile were uniform and equal to R we should have that if s is the short distance moved,

$R \cdot s =$ Work done

$= $ K.E. possessed just after impact by tup and pile

$$= \frac{(W + W_1) v^2}{2g}$$

$$= \frac{(W + W_1) \cdot W_1{}^2 v_1{}^2}{2g \cdot (W + W_1)^2} = \frac{W_1{}^2 h}{W + W_1}$$

$$= \frac{W_1 h}{\left(\dfrac{W}{W_1} + 1\right)} \; ;$$

$$\therefore R = \frac{W_1 h}{s \left(1 + \dfrac{W}{W_1}\right)} \quad \ldots (1).$$

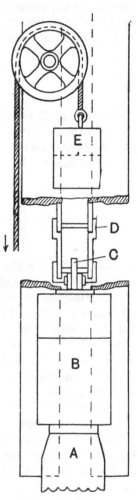

Fig. 74.   Pile-Driver.

This formula is not, however, strictly applicable to this problem because the resistance is not uniform.

A formula which is used a good deal in practice in order to determine the safe load P to put upon a pile is

$$P = \frac{2 W_1 h}{x + 1} \quad \ldots \ldots (2),$$

where h is the drop of the tup in feet,

and x is the penetration in inches of the last blow.

This formula professes to give a safe load on the pile equal to $\frac{1}{6}$ of the average resistance of the last blow.

That is $P = \dfrac{R}{6}$ of our formula (1).

Putting therefore $R = 6P$,

we have $$P = \frac{W_1 h}{6s \left(1 + \dfrac{W}{W_1}\right)},$$

or if h is in feet and for s which is also in feet we write $\dfrac{x}{12}$ where x is in inches we shall have

$$P = \frac{W_1 h}{\dfrac{x}{2}\left(1 + \dfrac{W}{W_1}\right)}$$

$$= \frac{2W_1 h}{x\left(1 + \dfrac{W}{W_1}\right)}$$

$$= \frac{2W_1 h}{x + \dfrac{W}{W_1}x} \quad \dots\dots\dots\dots(3).$$

If, therefore, $\dfrac{W}{W_1} \cdot x$ is equal to 1, the formula used in practice is equivalent to that which we have obtained from theoretical considerations which are not strictly applicable.

SUMMARY OF CHAPTER VIII.

Newton's Laws of Motion.

(1) A body continues in a state of rest or uniform motion in a straight line unless it be acted upon by some external force.

(2) The rate of change of momentum is proportional to the force applied and takes place in the direction of the force.

(3) To every action there is an equal and opposite reaction.

Suddenly applied forces are called *impulsive forces.*

$$F = \frac{Wv}{gt}.$$

Momentum before and after impact is equal.

Although no momentum is lost in impact there is always a loss of energy.

EXERCISES.   VIII.

1. A hammer head weighing $2\frac{1}{2}$ lbs. and moving with a velocity of 50 ft. per sec. is stopped in ·001 second. What is the average force of the blow?

2. A ship weighing 2000 tons and moving with a speed of 3 knots is stopped in 1 minute. Neglecting the motion of the water find the average retarding force if 1 knot is 6080 feet per hour.

3. A cage weighing 1000 lbs. is being lowered down a mine by a cable. Find the tension in the cable (1) when the speed is increasing at the rate of 5 feet per second per second; (2) when the speed is uniform; (3) when the speed is diminishing at the rate of 5 feet per second per second. The weight of the cable itself may be neglected.

4. A jet of water 1 inch in diameter falling from a height of 200 feet strikes a fixed hemispherical cup so as to reverse its direction. Find the force which it exerts upon the cup assuming that the jet has 90 per cent. of the full velocity due to its height of fall.

5. A gun delivers 400 bullets per minute, each weighing ·5 oz., with 2000 feet per second horizontal velocity; neglecting the momentum of the gases, what is the average force exerted upon the gun?

6. A 1 oz. bullet fired horizontally with a velocity 1000 feet per second into a 1 lb. block of wood resting on a smooth table penetrates 2 inches and remains embedded. With what velocity does the block move off? Would the bullet have penetrated more or less if the block had been fixed?

7. An 1800 lb. shot moving with a velocity of 2000 feet per second impinges on a plate weighing 10 tons, passes through it and goes on with a velocity of 400 feet per second. If the plate is free to move find its velocity.

8. Two inelastic bodies moving in the same direction with velocities of 10 and 8 feet per second impinge. If they weigh 4 and 5 lbs. respectively, what is their common velocity after impact? What would have happened if they had been moving in opposite directions?

CHAPTER IX

STRESS AND STRAIN

Strain may be defined as the change in shape or form of a body caused by the application of external forces.

Stress may be defined as the force between the molecules of a body brought into play by the strain.

An elastic body is one in which for a given strain there is always induced a definite stress, the stress and strain being independent of the duration of the external force causing them, and disappearing when such force is removed. A body in which the strain does not disappear when the force is removed is said to have a *permanent set* and such body is called a *plastic body*.

When an elastic body is in equilibrium the resultant of all the stresses over any given section of the body must neutralise all the external forces acting over that section. When the external forces are applied, the body becomes in a state of strain, and such strain increases until the stresses induced by it are sufficient to neutralise the external forces.

For a substance to be useful as a material of construction, it must be elastic within the limits of the strain to which it will be subjected. Most solid materials are elastic to some extent, and after a certain strain is exceeded they become plastic.

Hooke's Law—enunciated by Hooke in 1676—states that in an elastic body the *strain is proportional to the stress*. Thus, according to this law, if it take a certain weight to stretch a rod a given amount, it will take twice that weight to stretch the rod twice that amount; if a certain weight is required to make a beam deflect to a given extent, it will take twice that weight to deflect the beam to twice that extent.

Kinds of Strain and Stress. Strains may be divided into three kinds, viz. (1) an *extension*; (2) a *compression*; (3) a *slide*.

Corresponding to these strains we have (1) *tensile* stress; (2) *compressive* stress; (3) *shear* stress.

A body that is subjected to only one of these, is said to be in a state of *simple* strain, while if it is subjected to more than one, it is said to be in a state of *complex* strain.

Examples of simple strains are to be found in the cases of a tie bar; a column with a central load; a rivet, Fig. 75. The best example of a body under complex strain is that of a beam in which, as we shall show later, there exist all the kinds of strain.

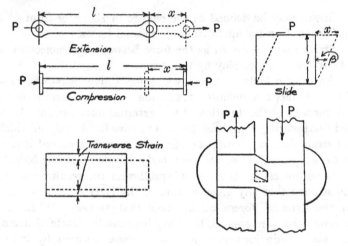

Fig. 75.   Kinds of Strain.

Intensity of stress.   Imagine a small area a situated at a point X in the cross section of a body under strain, then if S is the resultant of all the molecular forces across the small area, $\frac{S}{a}$ is called the *intensity of stress* at the point X.   In the case of bodies under complex strain, the intensity of stress will be different at different points of the cross section, while in a body subjected to a simple strain, the stress will be the same at each point of the cross section, so that in this case if A is the area of the whole cross section and P is the whole force acting over the cross section, the intensity of stress will be equal to $\frac{P}{A}$.   In future, unless it is stated to the contrary, we shall use the word "stress" to mean the "intensity of stress."

Unital strain. The unital strain is the strain per unit length of the material. In the case of extension and compression, the total strain is proportional to the original length of the body. Thus, a rod 2 ft. long will stretch twice as much as a rod 1 ft. long for the same load. In Fig. 75 if l is the unstrained length of the rods under tension and compression and x the extension or compression, the unital strain is $\frac{x}{l}$.

In the case of slide strain, the angle of the unit cube (Fig. 75) under consideration but not the length of the body is altered, and this angle β is the measure of the unital strain. If the angle is small, as it always will be in practice with materials of construction, then it will be nearly equal to $\frac{x}{l}$, where x and l are the quantities shown in the figure.

Stress-strain Diagrams. If a material be tested in tension or compression, and the strain at each stress be measured, and such strains be plotted on a diagram against the stresses, a diagram called the *stress-strain diagram* is obtained. If a material obeys Hooke's Law, this diagram will be a straight line. For most metals, the stress-strain diagram will be a straight line until a certain point is reached, called the *elastic limit*, after which the strain increases more quickly than the stress, until a point called the *yield point* is reached, where there is a sudden comparatively large increase in strain. After the yield point is reached, the metal becomes in a plastic state and the strains go on increasing rapidly until fracture occurs.

Fig. 76 shows the stress-strain diagram for a tension specimen of mild steel, such as is suitable for structural work.

The portion AB of the diagram is a straight line, and represents the period over which the material obeys Hooke's Law. At the point C, the yield point is reached, and the strain then increases to such an extent that the first portion of the diagram is re-drawn to a considerably smaller scale, as shown on the left in the figure. The strain then increases in the form shown until the point D is reached, the curve between C and D being approximately parabolic in shape. When the point D is reached, the maximum stress has been reached, and the specimen begins to

pull out and thin down at one section, and if the stress is
sustained, fracture will then occur. The portion *DE*, shown
dotted, represents increase of strain with apparent diminution of
stress. This diminution is only apparent because the area of the
specimen beyond the point rapidly gets smaller, so that the *load*
may be decreased and still keep the *stress* the same. In practice
it is very difficult to diminish the load so as to keep pace with
the decrease in area, so that this last portion of the curve is very
seldom accurate, and has, moreover, little practical importance.

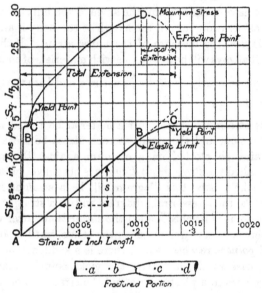

Fig. 76. Stress-strain Diagram for Mild Steel in tension.

The specimen draws down at the point of fracture in the
manner shown in the diagram. Before the test, it is customary
to make centre-punch marks at equal distances apart along the
length of the specimen. The distance apart of these points after
the fracture of the specimen indicates the distribution of the
elongation at different points along the length. Four such
marks, *a*, *b*, *c*, *d*, are shown in the figure. The greatest extension
occurs at the point of fracture, so that with a specimen of short
length, the percentage total extension will be greater than with
a longer specimen.

The stress-strain diagrams in compression and shear for mild steel are very similar to that for tension. In compression it is difficult to get the whole diagram, because failure occurs by *buckling*, except with very short lengths, where it is very difficult to measure the strains, and in shear the test has to be made by torsion, because it is almost impossible to eliminate the bending effect. Now, in torsion, the shear stress is not uniform, so that the metal at the exterior of the round bar reaches its yield point before the material in the centre, and this has the effect of ráising the apparent yield point. The same occurs in testing for compression or tension by means of beams.

The importance of the elastic limit has been overlooked to a great extent by designers of structures and machines; but inasmuch as the theory, on which most of the formulae for obtaining the strength of beams are based, assumes that the stress is proportional to the strain, it must be remembered that our calculations are true only so long as Hooke's Law is true, so that the elastic limit of the material is a very important quantity.

Stress-strain Diagrams for Cast Iron. The strength of cast iron varies largely with the composition, and the strength in tension

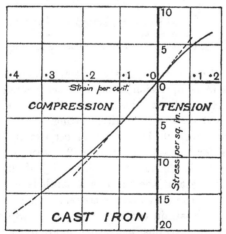

Fig. 77. Stress-strain Diagrams for Cast Iron.

is considerably less than that in compression. Fig. 77 shows the stress-strain diagrams for both tension and compression. It will be seen that in tension the strain is never really proportional to the stress, while in compression the stress and strain are approximately proportional up to a stress of about 8 tons per square inch. In the figure the compression curve is not completed, owing to buckling setting in. It is on account of the fact that the strain is not proportional to the stress that there is a considerable difference between the actual and calculated strengths of cast iron beams.

Other Materials. *Timber.*—There are several difficulties attendant upon the accurate testing of timber, owing to the effect of dampness and to lack of homogeneity in the material. It may be taken that the stress-strain diagrams are approximately straight for a portion of their length, but then curve off in a similar manner to the compression curve for cast iron.

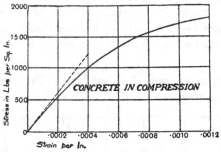

Fig. 78. Stress-strain Diagram for Concrete in compression.

Cement and Concrete.—The stress-strain diagram for cement and concrete in compression is never exactly straight, so that there is no elastic limit, the exact curve depending on the composition and on the time after setting.

The curve shown in Fig. 78 is almost exactly a parabola. This curve is for a 1—3—6 concrete, 90 days old, which was tested by Mr R. H. Slocom of the University of Illinois. Some authorities assume that the curve is a parabola but in practice it is seldom that the curve comes so near to a parabola as the above. The stress-strain curve is, however, nearly always of a similar shape, the strains increasing more quickly than the stresses. It is extremely important to remember that with cement and concrete the relations between stress and strain vary largely with the quality and proportions of ingredients, and cannot be taken as almost constant as in the case of steel. In tension a somewhat similar curve is obtained, but as cement and concrete are practically never used in tension, much less work has been done on its tensile strength.

The Elastic Constants or Moduli. If a material is truly elastic, i.e. if the strain is proportional to the stress, then it follows that the intensity of stress is always a certain number of times the unital strain, or that the ratio $\dfrac{\text{intensity of stress}}{\text{unital strain}}$ is constant. This stress-strain ratio is called a *modulus*. That for tension and compression is generally known as *Young's modulus*, and is given the letter E; that for shear is called the *shear* or *rigidity modulus* (G). There is an additional modulus called the *bulk* or *volume modulus* (K) which represents the ratio between the unital change in volume and the intensity of pressure or tension on a cube of material subjected to pressure or tension on all faces.

Young's modulus is the one which we shall be most concerned with in engineering design. Suppose a tension member (a *tie* as it is called) or a compression member (a *strut*) of length l and cross sectional area A is subjected to a pull or thrust P, and that the extension or compression is x, Fig. 75. Then the intensity of stress is $\dfrac{P}{A}$, and the unital strain is $\dfrac{x}{l}$.

$$\therefore \text{ Young's modulus} = E = \frac{P}{A} \div \frac{x}{l} = \frac{Pl}{Ax}.$$

The value of Young's modulus can be found from the stress-strain diagram. Thus in that for mild steel, Fig. 76,

$$E = \frac{s}{x}.$$

Now in the relation $E = \dfrac{\text{stress}}{\text{strain}}$, if the strain is equal to 1, i.e. if the bar is pulled to twice its original length, we have that $E = $ stress, and this accounts for the definition of Young's modulus that some writers have given, viz. "Young's modulus is the stress that is necessary to pull a body to twice its original length." Some students find this definition more clear than the one previously given, but it must be remembered that no material of construction will pull out to twice its original length without fracture.

Numerical Example. *A mild steel tie-bar, 12 ins. long and of $1\frac{1}{2}$ ins. diameter, is subjected to a pull of 18 tons. If the extension is ·0094 in., find Young's modulus.*

Area of section of $1\frac{1}{2}$ ins. diam. = 1·767 sq. ins.;

$$\therefore \text{Stress per sq. in.} = \frac{18}{1 \cdot 767} = 10 \cdot 19 \text{ tons per sq. in.}$$

$$\text{Unital strain} = \frac{\cdot 0094}{12} = \cdot 000783;$$

$$\therefore \text{Young's modulus} = \frac{10 \cdot 19}{\cdot 000783} = \underline{13,000 \text{ tons per sq. in.}}$$

Young's modulus for Concrete and similar Substances. If Young's modulus is a constant, it can be found for strains and stresses below the elastic limit only, and, strictly speaking, there is no modulus for substances such as concrete, where the strain is not proportional to the stress. From Fig. 78 it is clear that since the strain increases more quickly than the stress in concrete, the value of the ratio $\dfrac{\text{stress}}{\text{strain}}$ will be greater for small stresses than for large stresses, and so, before the value of this ratio is of any use to us, we must know the value of the stress at which the ratio is calculated. One can hardly lay too great stress on the importance of having exact ideas on the principles which form the foundations on which the theory of structures is built, and with concrete it is practically useless to speak of the compressive strength and Young's modulus unless the composition of the concrete and the stress at which the modulus is calculated are known.

Poisson's Ratio—Transverse Strain. When a body is extended or compressed, there is a transverse strain tending to prevent change of volume of the body. The amount of transverse strain bears a certain ratio to the longitudinal strain.

$$\text{This ratio} = \frac{\text{transverse strain}}{\text{longitudinal strain}} = \eta \text{ varies from } \tfrac{1}{3} \text{ to } \tfrac{1}{4} \text{ for}$$

most materials, and is called *Poisson's ratio.*

According to one school of elasticians, the value of this ratio η should be $\frac{1}{4}$, but experimental evidence does not quite support this view, although it is very nearly true for some materials. The ratio is very difficult to measure directly.

Experiment upon the tensile strength of wire. Although it requires a heavy testing machine to make some experiments upon the strength of materials, the following comparatively simple apparatus will enable a good deal to be learnt concerning the tensile strength of wire.

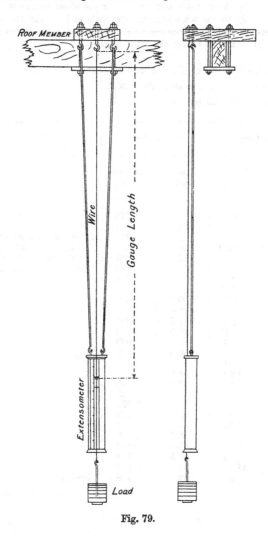

Fig. 79.

A wire about 8 feet long is suspended from the ceiling or other convenient point as shown in Fig. 79, and a scale pan is hung on the end. Suspended from the same point is a strain-measuring device or extensometer constructed

10—2

conveniently as shown in Fig. 80. A white celluloid vernier is carried by
a slider which is clipped to the wire at a fixed length l from the point of
suspension, this length being called the gauge length. By means of this vernier,
the length l for different loads upon the wire can be calculated. The following
records of an experiment made with this apparatus illustrate its use. Weights
are added gently to the scale pan and the scale and vernier are read after
each addition. The load is then the weight of the scale pan + the total weight
added as will be followed from the third column:

MATERIAL—SOFT IRON. GAUGE LENGTH—84".

Diam. (mean of six measurements) ·0364".

Weight of scale pan 0·5 lb.

Observed		Calculated		
Added wts.	Scale reading	Load lbs.	Extension ins.	
3	·65	3·5	·013	*NOTES*
3	·66	6·5	·023	
2	·67	8·5	·033	Extension is scale reading less
2	·675	10·5	·038	·637, the correct zero as
2	·68	12·5	·043	determined from the pre-
2	·685	14·5	.048	liminary plotting (Fig. 81.)
2	·69	16·5	·053	
2	·695	18·5	·058	
2	·70	20·5	·063	
2	·71	22·5	·073	
2	·72	24·5	·083	
2	·725	26·5	·088	
2	·73	28·5	·093	
2	·74	30·5	·103	
2	·84	32·5	·113	
2	1·40	34·5	·763	
2	2·20	36·5	1·563	
2	2·65	38·5	2·013	
2	3·23	40·5	2·593	
2	3·9	42·5	3·263	
2	4·65	44·5	4·013	
2	5·6	46·5	4·963	
2	6·8	48·5	6·163	
2	8·45	50·5	7·813	
2	11·2	52·5	10·563	
1	13·8	53·5	13·163	Broke.

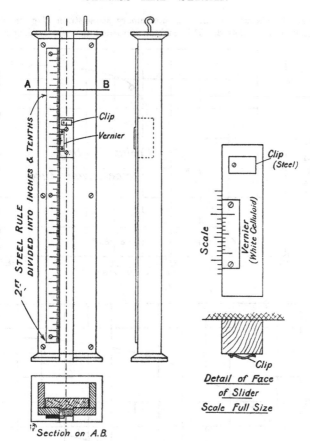

Fig. 80.

RESULTS.

Initial Area = ·785 × (·0364)² = ·00104 sq. in.

Final Area = ·785 × (·0334)² = ·000876 sq. in.

$$E = \frac{W}{a} \times \frac{l}{\Delta} = \frac{28}{·0925} \times \frac{84}{·00104} = \underline{24,400,000 \text{ lbs. per sq. in.}}$$

$$\text{Breaking Stress} = \frac{53·5}{·00104} = 51,400 \text{ lbs. per sq. in.}$$

$$= \underline{22·9 \text{ tons per sq. in.}}$$

$$\text{Stress at Elastic limit} = \frac{29}{·00104} = 27,900 \text{ lbs. per sq. in.}$$

$$= \underline{12·4 \text{ tons per sq. in.}}$$

$$\frac{\text{Elastic limit Stress}}{\text{Breaking Stress}} = \frac{29}{53·5} = \underline{·54.}$$

$$\text{Extension} = \frac{13·16}{84} = \underline{15·7 \text{ per cent.}}$$

From the observed results the preliminary stress-strain diagram shown in Fig. 81 is first plotted. This enables us to find the zero reading on the

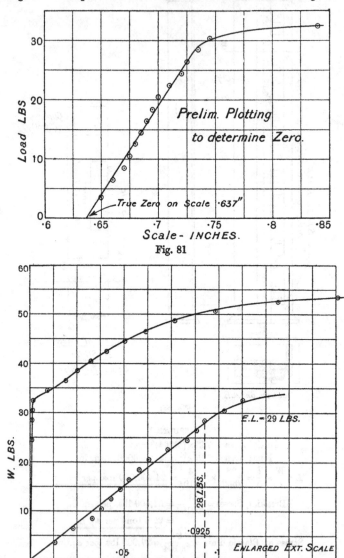

Fig. 81

Fig. 82.

scale from which we are able to calculate the extensions given in the fourth column and thus to plot the complete stress-strain diagram shown in Fig. 82.

This is strictly a diagram of loads plotted against extensions and is not one of intensity of stress plotted against unital strains. The real stress-strain diagram will have the same shape but a different scale so long as the area remains constant. In practice the breaking stress is always calculated by dividing the breaking load by the original area.

Factor of Safety. We will use the term "factor of safety" to denote the ratio $\dfrac{\text{Breaking stress of material}}{\text{Working stress used in design}}$. This is often taken the same as $\dfrac{\text{Load required to cause failure}}{\text{Load carried}}$. The two results are not the same because the theories that we use in design do not allow for all the possible contingencies. When using the second definition it is usual to specify a factor of safety of 4 for structures not subjected to impulsive loads and vibrations.

The term "factor of safety" is used very loosely in practice and it is very desirable to use it in some definite sense as the first one given above.

The great difficulty of getting a satisfactory definition of the factor of safety resides in the fact that we can find the load to require failure of a structure only by means of a test to destruction which is too costly and defeats its own end. In machine parts the same difficulty does not arise because the cost of a part to be tested will be small; the difficulty there is that the part must be tested under the actual conditions in which it is used in practice. For instance it is no use testing motor car axles to get an accurate value of the factor of safety by making a tension test on a piece of the material. Since the elastic limit in steel is the point which determines the real safety of a structure it would be more satis- factory to define the factor of safety in steel and other metals possessing an elastic limit as the ratio

$$\frac{\text{Maximum calculated stress in material}}{\text{Elastic limit stress in material}}.$$

Working Stresses. In the absence of other regulations it is usual to take the working stresses as $\frac{1}{4}$ of the breaking stresses (see table on p. 157) but for cast iron $\frac{1}{8}$ is often taken. For mild steel, wrought iron and cast iron the following values may be safely taken where the body is not subject to shocks and vibra- tions.

	Working Stresses in tons per sq. in.		
	Tension f_t	Compression f_c	Shear f_s
Mild steel	7	6	5
Wrought iron	5	4	4
Cast iron	1	6	1

Work done in Straining; Resilience. If we look at a stress-strain diagram, we shall see that the strains are, to a reduced scale, the total distance moved by the end of the specimen because extension = distance moved = unital strain × original length. Also in simple stressing we have

$$\text{Stress} = f = \frac{\text{Load}}{\text{Area}} = \frac{F}{A}.$$

∴ $F = Af$, but F is equivalent to the force that we have previously spoken of as the *effort*, so that if the area keeps constant—as it does for all practical purposes within the elastic limit—the stress is a measure of the effort.

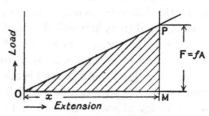

Fig. 83.   Resilience.

The stress-strain diagram therefore is a special kind of diagram of effort plotted against distance and we have previously called such a diagram the *effort curve* (p. 43).

We have proved on p. 43 that the work done is equal to the area below the effort curve; therefore the area below the load-extension diagram must give the work done in straining the bar. Now the *work done in straining a material per unit volume of the material is called the Resilience.*

Referring to Fig. 83, let l be the original length of the bar.

The work done in producing a stress f

$$= \text{Area of } \triangle POM$$
$$= \tfrac{1}{2} PM . MO$$
$$= \tfrac{1}{2} F . x$$
$$= \tfrac{1}{2} f . A . x.$$

Now $\dfrac{x}{l}$ = unital strain = $\dfrac{\text{stress}}{E}$ = $\dfrac{f}{E}$;

$$\therefore x = \frac{fl}{E};$$

$\therefore$ Work done in producing a stress f

$$= \tfrac{1}{2} . f . A . \frac{fl}{E}$$
$$= \frac{f^2}{2E} \times Al.$$

But Al = volume of the bar = V,

$$\therefore \frac{\text{Work}}{\text{Volume}} = \text{Resilience} = \frac{f^2}{2E}.$$

Stresses and Strains due to Sudden or Dynamic Loading.
If a load is applied suddenly to a machine or structure, vibration
will ensue, and the strain—and thus the stress—will reach twice
the value which would occur if the load were gradually applied.

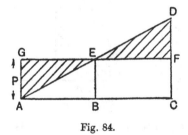

Fig. 84.

This will be made clear from considering a diagram, Fig. 84,
where the force is plotted against the strain. We have seen that,
with gradual loading of an elastic body, the curve representing
the relation between the strain and the load in direct stress is
represented by a straight line AD, the area below the line giving
the work done up to a given point. Now let AG represent a
force P; then when the strain gets to the point B, the work

done by the force will be equal to the area of the rectangle *ABEG*, whereas the work done in straining the material is only equal to the area of the triangle *ABE*, so that there is an amount of work equal to the area of the triangle *AEG* still available for causing increased strain. The strain therefore increases until the area of the triangle *EFD* is equal to that of the triangle *AEG*. It is clear that $AC = 2AB$, or *that the strain—and thus the stress —is twice that in the case of gradual loading.*

This is a most important point and shows us that we should make allowance in engineering calculations for the nature of the loading, whether gradual or sudden. In the latter case therefore we ought to allow a greater factor of safety.

Gradual loads are usually called "static loads" or "dead loads," and sudden loads are called "dynamic loads" or "live loads."

It is a good rough rule to take a live load as equivalent to a dead load of twice its value.

Strain and Stress due to Impact. Suppose a weight W falls from a height h on to a structure and let the deformation or strain in the direction of h be x, Fig. 85. Then the work done

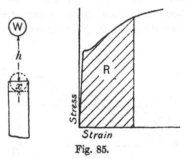

Fig. 85.

by the weight is equal to $W(h + x)$. Now this work is absorbed in straining the structure. Consider first the case in which the resulting strain is within the elastic limit. The work done in such case is equal to the volume multiplied by the resilience. We have shown that in tension or compression the resilience is equal to $\dfrac{f^2}{2E}$ and therefore in this case we get

$$W(h + x) = \frac{\text{Volume} \times f^2}{2E} = \frac{Vf^2}{2E}.$$

Then if x is negligible compared with h we have

$$W \times h = \frac{Vf^2}{2E},$$

or
$$f = \sqrt{\frac{2EWh}{V}}.$$

If the weight strikes with a velocity v,

$$h = \frac{v^2}{2g},$$

or
$$f = \sqrt{\frac{2E \cdot Wv^2}{2gV}} = v\sqrt{\frac{EW}{gV}}.$$

Strain beyond Elastic Limit. If the strain is beyond the elastic limit, it follows, from the reasoning given on p. 153, that the work done per unit volume in straining is equal to the area below the stress-strain curve. If this area is R, Fig. 85, then we have $R = Wh$ or $\frac{Wv^2}{2g}$.

From this the stress can be found.

Numerical Example. *A bar of $\frac{1}{2}$ inch diameter stretches $\frac{1}{8}$ inch under a steady load of 1 ton. What stress would be produced in the bar by a weight of 150 lbs. which falls through 3 inches before commencing to stretch the bar—the bar being initially unstressed and the value of E taken as 30×10^6 lbs. per square inch?*

Area of bar $\frac{1}{2}''$ diam. $= \cdot 196$ sq. in.

$\therefore$ Stress under load of one ton $= \dfrac{1}{\cdot 196}$ tons per sq. in.

$$= \frac{2240}{\cdot 196} \text{ lbs. per sq. in.}$$

$$\therefore \text{Strain} = \frac{\text{Stress}}{E} = \frac{2240}{\cdot 196 \times 30 \times 10^6}.$$

Now $\frac{1}{8}'' = $ Strain $\times$ Original length,

$$\therefore \text{Original length} = \frac{\frac{1}{8}}{\text{Strain}} = \frac{\cdot 196 \times 30 \times 10^6}{2240 \times 8};$$

$$\therefore \text{Volume} = \text{Length} \times \text{Area of section}$$

$$= \frac{\cdot 196 \times \cdot 196 \times 30 \times 10^6}{8 \times 2240}$$

$$= 64 \cdot 31 \text{ cub. ins.}$$

Work done by 150 lbs. in falling 3 inches $= 3 \times 150 = 450$ in.-lbs.

$$\therefore \frac{64 \cdot 31 \times f^2}{2E} = 450,$$

$$f = \sqrt{\frac{900E}{64 \cdot 31}}$$

$$= \sqrt{\frac{900 \times 30 \times 10^6}{64 \cdot 31}}$$

$$= \underline{20,490 \text{ lbs. per sq. in.}} \quad Ans$$

Temperature Stresses. Suppose a bar of length l is heated $t°$ F. and a is coefficient of expansion. Then, unless prevented, the length of the bar will become $l(1 + at)$, i.e. the increase in length will be atl.

If the bar is rigidly fixed so that this expansion cannot take place, then there will be in the bar a strain equal to atl, and the unital strain will be $\frac{atl}{l} = at$.

This strain will produce a compressive stress of $at \times E$, where E is Young's modulus.

Now for mild steel $a = \cdot 00000657$ per degree Fahrenheit, and $E = 13,000$ tons per square inch.

$$\therefore \text{ The stress per } ° \text{ F.} = \cdot 00000657 \times 13,000$$

$$= \cdot 0854 \text{ ton per square inch.}$$

Taking a range of temperature of 120° F., the stress due to temperature $= 120 \times \cdot 0854 = 10 \cdot 25$ tons per square inch. This is more than the safe stress for mild steel, so that the importance of designing structures so that the expansion may take place becomes quite evident.

Struts, Columns and Pillars. When bars are in compression they are called struts, columns or pillars; under test such bars always fail by *buckling* as indicated in Fig. 86, buckling being a kind of bending. Full consideration of this question is rather difficult and beyond our present scope. We will just point out the following facts:

(1) The strength of a strut depends upon its length and shape of cross section as well as upon its cross sectional area.

ELASTIC PROPERTIES OF MATERIALS.

Material	Weight per cub. foot (lbs.)	Breaking Stress			Elastic Moduli		Elastic Limit
		Tension	Crushing	Shear	E	C	
Mild steel	490	28–32	—	20–25	13,000	5200	15–20
Wrought iron	480	20–25	16–22	16–19	12,500	5000	10–16
A ⎰ Cast iron	440–470	5–15	25–65	6–13	5000–8000	2500–4000	—
Copper	540	12–15	20–25	11–12	7000	—	3–8
Gun-metal	,,	,,	25–40	9–12	5000	—	5–7
Brass	520–530	10–12	—	—	6000	—	—
Timber	30–50	1½–7	2–4	¼–1	600–1000	—	—
Portland cement	90	500–700	4000–6000	—		—	—
Gravel concrete (1 : 2 : 4)	120	250	2500	—	2×10^{6} to 5×10^{6}	—	—
,, ,, (1 : 3 : 6)	130	—	2000	—		—	—
Cinder concrete	115	200	500	—	—	—	—
B ⎰ Brick (London stock)	140	—	2000–3000	—	—	—	—
,, (Staffordshire blue)	100–150	—	7000	—	—	—	—
Brickwork in cement		—	1250–2500	—	—	—	—
Portland stone	145	—	5000	—	—	—	—
Sandstone	135–145	—	5000–10,000	—	—	—	—
Granite	170	—	12,000–23,000	—	—	—	—

The stresses, etc., for materials A are in tons per square inch, and for materials B are in lbs. per square inch.

(2) In choosing a column section we should try to get it about as broad one way as the other and as much of the material should be as far as possible from the centre. For the latter reason, shapes (*b*) and (*d*) are better than (*a*) and (*c*) but all are fairly suitable.

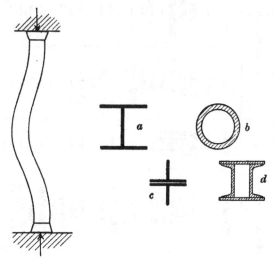

Fig 86.

SUMMARY OF CHAPTER IX.

Strain is the change in shape or form of a body caused by the application of external forces.

Stress is the force between the molecules of a body brought into play by the strain.

Hooke's Law states that in an elastic body strain is proportional to the stress.

There are three kinds of stress: tension, compression and shear.

Intensity of stress is the total stress on a small area divided by the area.

Unital strain is the strain per unit length of the material.

The *elastic limit* of a material is the stress at which the strain ceases to be proportional to the stress.

The *yield point* is the stress at which the strain increases suddenly without increase of stress.

Young's modulus. $E = \dfrac{\text{Tensile or compressive intensity of stress}}{\text{Tensile or compressive unital strain}}$

$$= \frac{Pl}{Ax}.$$

Factor of safety $= \dfrac{\text{Breaking stress}}{\text{Working stress used in design}}$.

Poisson's ratio $= \dfrac{\text{Transverse strain}}{\text{Longitudinal strain}}$.

Resilience is the work done in straining a material per unit volume.

Gradually applied loads are called *dead loads*; suddenly applied ones are called *live loads*.

A live load causes twice the stress caused by a dead load of the same amount.

EXERCISES. IX.

1. A tie-rod in a roof whose length is 142 ft. stretches 1 inch when bearing its proper stress. What strain is it subjected to?

2. How much will a tie-rod 100 ft. long stretch when subjected to ·001 of strain?

3. A cast iron pillar 18 ft. high shrinks to 17·99 ft. when loaded. What is the strain?

4. A tie-rod 100 ft. long has a sectional area of 2 sq. ins., it bears a tension of 32,000 lbs. by which it is stretched ¾″. Find the intensity of the stress, the strain, and the modulus of elasticity.

5. How much will a steel rod 50 ft. long and ⅛ sq. in. sectional area be stretched by a weight of one ton, the modulus of elasticity being 35,000,000 lbs. per sq. in.?

6. Find the work done in stretching each of the rods in Questions 4 and 5.

7. The diameter of the piston of an engine is 12″, the diameter of the piston-rod being 2¼″. Find the stress in the piston-rod when the maximum steam pressure is 120 lbs. per sq. in.

8. Find the proper diameter for a wrought iron rod to sustain a direct pull of 13 tons, the greatest stress allowable being 9000 lbs per sq. in.

9. The diameter of a steel rod is 2 ins., find the greatest weight it could support so as not to stress it to more than 10,000 lbs. per sq. in.

10. What stress in lbs. per sq. in. will stretch an iron bar whose length is 12 ft. by ¼ in.? The modulus of elasticity of iron being 28,000,000 lbs. per sq. in.

11. Define stress, strain and modulus of elasticity. The cross sectional area of a piece of wire is ·02 sq. in. and its length is 20 ft. When loaded with 150 lbs. it stretches ·08 in. Find the modulus of elasticity of the wire.

12. Find the work done in stretching the bar in Question 10.

13. An iron rod is suspended by one end. Draw a curve showing the stress at any section, and find the length of a rod which can just carry its own weight, allowing a working stress of 9000 lbs. per sq. in. and weight of material 480 lbs. per cubic ft.

14. The elastic limit of a bar was found to be 20,000 lbs. per sq. in. and the strain at this point was ·0006. What was the resilience of the material?

15. A load of 560 lbs. falls through ½ in. on to a stop at the lower end of a vertical bar 10 ft. long and 1 sq. in. in section. If $E = 13,000$ tons per sq. in. find the stress produced in the bar.

16. A tie-rod in a roof structure has to stand a total pull of 40 tons. If the breaking stress in the material is 30 tons per sq. in. and a factor of safety of 6 is required, find a suitable diameter for the bar.

17. What is Poisson's ratio? If a steel bar 3 inches in diameter and 6 inches long is subjected to an axial pull of 70 tons, find the longitudinal and transverse strains if $E = 30 \times 10^6$ lbs. per sq. in. and $\eta = \frac{1}{4}$.

18. A cylinder cover 10 inches in diameter is attached by 12 studs ⅝ inch diameter at the bottom of the thread (·3 square inch in area). Find the force per square inch in these, when the steam pressure is 100 lbs. per sq. in. by gauge.

CHAPTER X

RIVETED JOINTS; THIN CYLINDERS

Forms of Rivet Heads. The most common forms of rivet heads and their usual proportions are shown in Fig. 87.

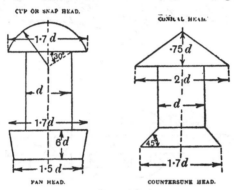

Fig. 87. Forms of Rivet Heads.

For structural work the snap-headed rivets are most usual, but countersunk rivets are used where necessary to prevent projections from the surface of the plate. Snap-heads take a length of rivet equal to about $1\frac{1}{4}$ times the diameter.

It is usual in practice to adopt a diameter of rivet when cold equal to one-sixteenth of an inch less than the diameter of the hole, but in all calculations the diameter of the rivet is taken as being equal to that of the hole.

Diameter of Rivets. According to Unwin's formula, the diameter of the rivet is $1\cdot2\sqrt{t}$, where t is the thickness of the thinnest plate, but for structural work this rule is very seldom adopted. In practice a $\frac{3}{4}''$ or $\frac{7}{8}''$ rivet is used wherever possible, and it is best not to use any formula to obtain the diameter in terms of the thickness of the plate. Some authorities use a diameter of $\frac{3}{4}''$ for a $\frac{3}{8}''$ plate, $\frac{7}{8}''$ for a $\frac{1}{2}''$ plate, and $1''$ for a $\frac{5}{8}''$ plate. It is difficult to get rivets of larger diameter than 1 in. driven by hand.

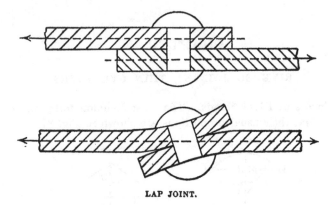

LAP JOINT.

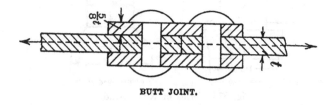

BUTT JOINT.

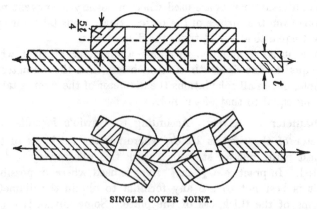

SINGLE COVER JOINT.

Fig. 88.

It is a mistake to adhere too rigidly to Unwin's formula; the
best diameter will be that which will give equal shearing and
bearing strengths (see p. 164 for explanation of these terms) and
will therefore depend upon whether the rivets are in double or
single shear. Basing our ideas upon this we should get the
following formulae:

$$d = 2 \cdot 5t \text{ for single shear}$$
$$= 1 \cdot 25t \text{ for double shear.}$$

For single shear this would make d so large for thick plates
that there would be practical difficulty in heading the rivets.

Forms of Joints. (a) LAP JOINTS AND BUTT JOINTS. In
the *lap joint* the plates overlap as shown in Fig. 88. This form
of joint has the disadvantage that the line of pull is such as to
cause bending stresses, tending to distort the joint as shown.

In the *butt joint* the edges of the plate come flush, and cover
plates are placed on each side as shown, the thickness of the
cover plates being each five-eighths that of the main plates. In
this form of joint the pull is central, so that there are no bending
stresses.

In the *single cover joint*, which is a cross between the lap joint
and the butt joint, there are bending stresses developed, tending
to distort the joint as shown.

It is clear from the above that the butt joint should be adopted
wherever possible.

(b) CHAIN RIVETING AND ZIG-ZAG OR STAGGERED RIVETING.

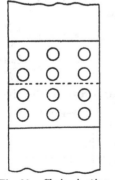   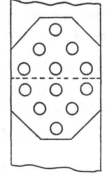

Fig. 89. Chain riveting.   Fig. 90. Zig-zag riveting.

The different rows of rivets in a joint may be arranged in chain
form or zig-zag form, as shown in Figs. 89, 90. As we shall see

11—2

later, the zig-zag form is more economical, and should be used
whenever possible.

Methods in which a Riveted Joint may Fail. A riveted joint
may fail in any of the following ways:

(1) By tearing of the plate.

(2) By shearing of the rivets.

(3) By crushing of the rivets.

(4) By bursting through the edge of the plate.

(5) By shearing of the plate.

Fig. 91 shows these methods of failure.

(4) and (5) are allowed for by the following rule: The minimum
distance between the centre of a rivet and the edge of the plate
is $1\frac{1}{2}d$, where d is the diameter of the rivet.

If this rule is adhered to the joint will always fail first in one
of the ways (1), (2), (3).

The aim in designing a joint should be to make the force
necessary to cause failure in the various ways equal.

We will now consider the various ways of failure in detail,
taking in each case a strip of plate equal to the pitch of the rivets.

(1) TEARING OF THE PLATE. In this case the width along
which fracture will occur is $(p - d)$, and as the thickness of the
plate is t, the area of fracture $= (p - d) t$.

Therefore if f_t is the *safe* tensile stress in the material, the safe
load which the joint can carry is equal to

$$P = f_t (p - d) t \quad \ldots\ldots\ldots\ldots\ldots(1).$$

(2) SHEARING OF THE RIVETS.

In the case of single shear, the area sheared $= \dfrac{\pi d^2}{4}$,

,,   ,, double   ,,   ,,   ,, $= \dfrac{2\pi d^2}{4}.$

Therefore if f_s is the safe shear stress on the rivet, the safe
orces on the joint as regards shear are respectively

$$\left. \begin{array}{l} P = f_s \dfrac{\pi d^2}{4} \\[2mm] P = f_s \dfrac{2\pi d^{2*}}{4} \end{array} \right\} \quad \ldots\ldots\ldots\ldots\ldots(2).$$

* A Board of Trade rule makes this 1·75 instead of 2, this being based
upon experiments. The rule is generally used in boilers but not so much in
bridge work.

(3) CRUSHING OR BEARING OF RIVETS. In this case the crushing or bearing area is taken as the diameter of rivet multiplied by the thickness of the plate, i.e. $d \times t$. Therefore, if f_B

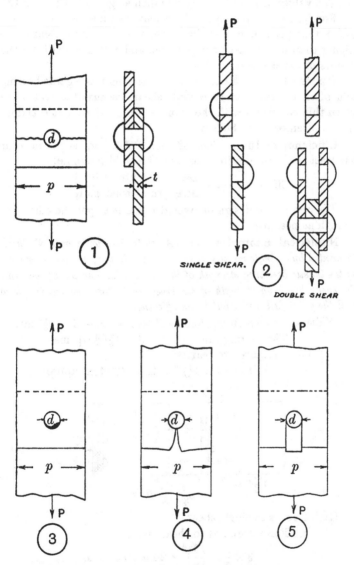

SINGLE SHEAR.

DOUBLE SHEAR

Fig. 91. Methods of failure of a Riveted Joint.

is the safe bearing stress on the rivet, the safe force on the joint
as regards bearing is equal to

$$P = f_B \cdot d \cdot t \quad \ldots\ldots\ldots\ldots\ldots\ldots (3).$$

The values of f_t and f_s may be taken as given in Chapter IX.

For f_B, 10 tons per square inch may be taken for mild steel,
and 8 tons per square inch for wrought iron. These figures are
higher than for ordinary compression, and are obtained from the
results of experiments.

For structural work the strength of the joint as regards bearing
will often be less than as regards shear, because the plates are
often thin compared with the diameter of the rivet, but this does
not so often occur in boilers.

Efficiency of Joint. The efficiency of a joint is the ratio of
the least strength of a joint to that of a solid plate, i.e.

$$\text{Efficiency} = \eta = \frac{\text{Least strength of joint}}{\text{Strength of solid plate}}.$$

The aim in the design of riveted joints is to get the efficiency
as high as possible.

Numerical Examples. (1) *A plate 10″ wide and ¾″ thick
is jointed by a single riveted lap joint with rivets of 1⅛ ins. diameter.
If the tensile breaking strength of the plate is 22 tons per sq. in. and
the shear breaking strength of the rivets is 20 tons per sq. in., how
will the joint fail and what is its efficiency?*

Width across which the joint will tear is $(10 - 3 \times 1\frac{1}{8})$ ins.

∴ The tearing area $= (10 - 3 \times 1\frac{1}{8}) \frac{3}{4}$ sq. ins.

∴ Force required to tear

$$= (10 - 3 \times 1\tfrac{1}{8}) \tfrac{3}{4} \times 22 = 109 \text{ tons approx.}$$

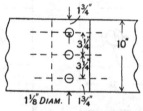

Fig. 92.

Each rivet is in single shear.

∴ Force required to shear rivets

$$= 3 \times \frac{\pi}{4} \times \left(\frac{9}{8}\right)^2 \times 20 = 59\text{·}5 \text{ tons.}$$

This is less than that for tearing.

∴ The joint will fail by shearing of the rivets.

$$\therefore \text{Efficiency} = \frac{\text{Least strength of joint}}{\text{Strength of original plate}}$$

$$= \frac{59\cdot5}{10 \times \tfrac{3}{4} \times 22} = \frac{59\cdot5}{165}$$

$$= \cdot361 = \underline{36\cdot1\,\%}.$$

(2) *Design a double-riveted lap joint to connect two steel plates $\tfrac{1}{2}$ in. thick with steel rivets. The tensile strength of the plates before drilling being 30 tons per sq. in.; the shearing strength of the rivets 24 tons per sq. in.; and the compressive strength of the steel 43 tons per sq. in. Find the efficiency of the joint.*

For $\tfrac{1}{2}$ in. plates Unwin's formula would give

$$d = 1\cdot2\sqrt{\cdot5} = \cdot85 \text{ in., say } \tfrac{7}{8} \text{ in.}$$

The joint is a double-riveted lap, therefore there will be two rivets in single shear in a width of plate equal to the pitch.

∴ Strength against tearing per pitch

$$= f_t\,(p-d)\,t$$
$$= 30\,(p-d)\,\tfrac{1}{2} = 15\,(p-d) \,..(1).$$

∴ Strength against shearing per pitch

$$= f_s \cdot \frac{2\pi d^2}{4}$$

$$= \frac{24\cdot2\pi}{4}\cdot\left(\frac{7}{8}\right)^2 \quad\ldots\ldots\ldots(2)$$

$$= 28\cdot9 \text{ tons.}$$

If these are equal,

$$15\left(p - \frac{7}{8}\right) = 28\cdot9\,;$$

$$\therefore p = \frac{28\cdot9}{15} + \frac{7}{8}$$

$$= 1\cdot93 + \cdot87 = 2\cdot80, \text{ say 3 ins.}$$

The bearing stress for a force of 29·8 tons would be equal to

$$\frac{28\cdot9}{\tfrac{7}{8} \times \tfrac{1}{2} \times 2} = 33 \text{ tons per sq. in.,}$$

the bearing area of each rivet being $\tfrac{7}{8} \times \tfrac{1}{2} = \cdot437$ sq. in.

This is less than the allowable value of 43 tons per sq. in., showing that a larger diameter of rivet might be used with greater economy, but $\frac{7}{8}$ in. diameter is in most cases more suitable in practice.

The efficiency of joint in this case is equal to

$$\frac{28\cdot9}{30 \times 3 \times \frac{1}{2}} = \frac{28\cdot9}{45} = \underline{64\cdot2\ \%}.$$

The joint is then as shown in Fig. 93.

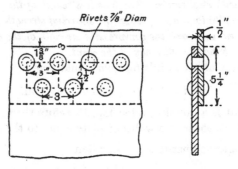

Fig. 93.

(3) *Find the number of rivets necessary to connect the gusset plates, etc., at the base of a steel stanchion to the stanchion proper, the load carried being 150 tons. The diameter of the rivets is $\frac{7}{8}$ in. and the thickness of the plate $\frac{1}{2}$ in.*

The rivets are best designed in such cases to carry the whole load, so that if the stanchion itself does not bear on the base plate the rivets will distribute the load satisfactorily.

The strength of each rivet in single shear

$$= \frac{\pi}{4} \cdot \left(\frac{7}{8}\right)^2 \cdot 5 = 3\cdot01 \text{ tons.}$$

The strength of each rivet in bearing

$$= \frac{7}{8} \cdot \frac{1}{2} \cdot 10 = 4\cdot37 \text{ tons.}$$

∴ Number of rivets necessary $= \dfrac{150}{3\cdot01} = \underline{50 \text{ nearly.}}$

WORKING STRENGTH OF STEEL RIVETS.

Diam. of Rivets in ins.	Area in sq. ins.	Strength in single shear at 5 tons per sq. in.	Bearing Strength at 10 tons per sq. in.							
			Thickness in ins. of plate							
			$\frac{5}{16}$	$\frac{3}{8}$	$\frac{7}{16}$	$\frac{1}{2}$	$\frac{9}{16}$	$\frac{5}{8}$	$\frac{11}{16}$	$\frac{3}{4}$
$\frac{3}{8}$	·1104	·55	1·17	1·41	1·64	1·87	2·11	2·34	2·59	2·81
$\frac{1}{2}$	·1963	·98	1·56	1·87	2·18	2·50	2·81	3·12	3·43	3·75
$\frac{5}{8}$	·3068	1·53	1·95	2·34	2·72	3·12	3·51	3·90	4·30	4·68
$\frac{3}{4}$	·4418	2·21	2·34	2·81	3·27	3·75	4·21	4·69	5·16	5·63
$\frac{7}{8}$	·6013	3·01	2·72	3·27	3·82	4·37	4·91	5·46	6·02	6·56
1	·7854	3·93	3·12	3·75	4·37	5·00	5·62	6·25	6·87	7·50

The Strength of Thin Cylinders and Pipes. Suppose that a thin pipe or cylinder such as is shown in Fig. 94 carries inside it a fluid such as air, steam or water under pressure. There are two principal ways in which it might fracture; firstly longitudinally as indicated at the bottom and secondly circumferentially at a line such as XX.

Longitudinal strength. We will first consider the stresses in a longitudinal section and will neglect the additional strength given by the two ends. Take for instance a length l of the pipe. The fluid under a pressure p lbs. per sq. in. acts radially all round the section as shown on the upper portion of diagram and the total pressure acting on the upper half of the length l will be equal to pressure × area $= \dfrac{\pi dl}{2} \times p$, but it is only the vertical component of the pressures which tends to cause the fracture under consideration. In the case of the pressure oa, for instance, ab is the part which is effective; we therefore get the result that the *effective total pressure tending to burst the pipe is given by pdl*, i.e. the force obtained by considering the pressure as acting upon the diameter.

If f_t is the tensile stress at the section and A is the area of fracture we shall therefore have

$$f_t \cdot A = pdl;$$

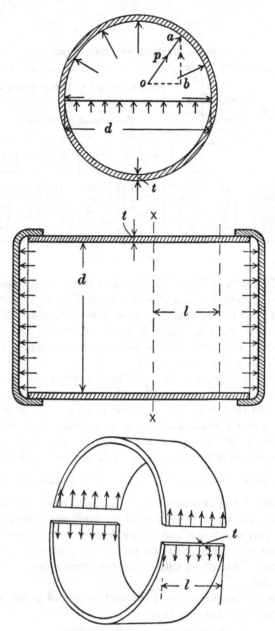

Fig. 94. Thin Cylinders and Pipes.

but the area of fracture is a rectangle at each side of length l and thickness t,

i.e. $$A = 2lt;$$
$$\therefore f_t \cdot 2lt = pdl,$$

i.e. $$f_t = \frac{pdl}{2lt} = \frac{pd}{2t} \quad \dots\dots\dots\dots\dots(1).$$

Circumferential Strength. The force tending to cause fracture in this case is the total bursting pressure upon the ends,

i.e. $p \times$ area of pipe $= \dfrac{p \times \pi d^2}{4}$.

The area of fracture is a thin circular ring of diameter d and thickness t; this area is for all practical purposes equal to circumference of inside of pipe × thickness of pipe = πdt;

$\therefore$ Stress × area of fracture = total bursting pressure,

i.e. $$f_t \times \pi dt = \frac{p \times \pi d^2}{4},$$
$$f_t = \frac{p \times \pi d^2}{4 \times \pi dt}$$
$$= \frac{pd}{4t} \quad \dots\dots\dots\dots\dots(2).$$

This is exactly one-half of the stress in the longitudinal section and shows that the longitudinal section is the weakest. In the design of boiler shells therefore the longitudinal seams are provided with stronger joints than the circumferential seams.

By means of formula (1) we can find the thickness required for a pipe of given diameter to withstand a certain pressure and keep the stress within certain fixed limits.

Numerical Example. *How thick would you make a boiler 6 feet in diameter which has to withstand a pressure of 200 lbs. per sq. in., if the stress in the material must not exceed 12,000 lbs. per sq. in. ?*

From equation (1)
$$f_t = \frac{pd}{2t},$$
$$f_t = 12,000, \quad p = 200, \quad d = 72 \text{ inches};$$
$$\therefore 12,000 = \frac{200 \times 72}{2t},$$
$$t = \frac{200 \times 72}{2 \times 12,000} = \cdot 6 \text{ inch,}$$
say ⅝ in.

SUMMARY OF CHAPTER X.

Riveted joints may be lap joints or butt joints.

Shear strength of a rivet

$$= \frac{f_s \pi d^2}{4} \text{ for single shear,}$$

$$\left. \begin{array}{l} = 1{\cdot}75 f_s \dfrac{\pi d^2}{4} \\[2mm] \text{or} \quad = 2 f_s \dfrac{\pi d^2}{4} \end{array} \right\} \text{ for double shear.}$$

Bearing strength of a rivet

$$= f_B . d . t.$$

$$\text{Efficiency of a riveted joint} = \frac{\text{Least strength of joint}}{\text{Strength of solid plate}}.$$

The circumferential strength of a thin cylinder is twice the longitudinal strength.

For longitudinal strength

$$f_t = \frac{pd}{2t}.$$

EXERCISES. X.

1. If the ultimate shearing strength of a steel plate is 20 tons per sq. inch, what force will be required to cause a 1 inch rivet in a single-riveted lap joint with a lap of 2 inches to shear through the plate which is $\frac{3}{4}$ in. thick?

2. What diameter of rivet would you use for a $\frac{1}{2}$ inch plate, and what pitch would you use for a lap joint? What would the efficiency of your joint be?

3. In a butt joint with a single row of rivets the plates are $\frac{1}{2}$ inch thick, the rivets are $\frac{7}{8}$ inch diameter and $1\frac{3}{8}$ inches apart; calculate the efficiency of the joint.

4. Two rectangular tie-bars are united by two cover-plates as shown (see Fig. X a). If $f_s = f_t = 6$ tons per sq. in., find the resistance to shearing of the rivets and to tearing of the plates.

What should be the thickness of the cover-plates? What is the efficiency of the joints?

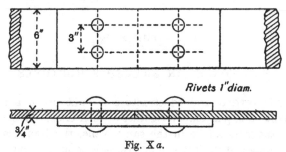

Rivets 1" diam.

Fig. X a.

5. How many rivets would you use to connect a member of a roof-truss to the main body of the truss if the member carries 20 tons and is ½ in. thick? The rivets are ⅞ in. diameter. Is bearing more important than shearing in this case?

6. What load may be safely carried by a column which has 40 rivets in single shear connecting the column to its base? The rivets are ⅞ in. diameter. Take a safe shear stress of 5 tons per sq. in.

7. For equal strengths in tension and shear calculate the pitch for a butt joint given the following data. Plates 1 inch thick; rivets 1¼ ins. diameter; two rows of rivets on each side of joint; $f_s = 54,000$ lbs. per sq. in.; $f_t = 65,000$ lbs. per sq. in.

8. A cylindrical boiler 8 feet in diameter is to withstand a working pressure of 100 lbs. per square inch. Calculate to the nearest ⅛ inch the thickness of the shell, allowing a stress of 10,000 lbs. per square inch, and neglecting the effect of the joint.

9. Find the thickness of an iron boiler shell to withstand an internal pressure of 250 lbs. per sq. in. The diameter is 10 ft. and safe stress allowable on plates 3 tons per sq. in.

10. What should be the thickness in the previous question if the joint has an efficiency of 60 per cent., the shell being composed of two plates?

11. Find the safe working pressure in lbs. per sq. in. for a boiler 6 ft. diameter, plates ½" thick; allowing a stress of 3500 lbs. per sq. inch.

12. The diameter of a steam boiler is 4 ft. Find the thickness of plates necessary if the stress is not to exceed 5 tons per sq. in. when the internal pressure is 120 lbs. per sq. in. and the efficiency of the joint is 72 per cent.

CHAPTER XI

THE FORCES IN FRAMED STRUCTURES

A THEORETICAL framed structure is built up of a number of straight bars, pin-jointed together at their ends. If the centre lines of the bars all lie in the same plane, the frame is called a *plane frame*; if in different planes, it is termed a *space frame*. For the present we shall deal only with plane frames.

A framed structure is designed so that, as far as possible, there are only pulling and thrusting forces, causing tension and compression stresses respectively, in its members, bending actions being obviated. In Continental and American practice it is common to make framed structures—or *trusses* as they are called—pin-jointed, but in British practice the joints are nearly always riveted. In either case the forces in the members, or *stresses* as they are usually rather erroneously called, are calculated as if the joints were pinned; these joints are often called *nodes*.

Kinds of Framed Structures. A framed structure may be one of three kinds, viz. deficient or under-firm, perfect or firm, and redundant or over-firm.

A *deficient or under-firm* frame is one which has not sufficient bars to keep it in equilibrium for all systems of loading. Such a frame is shown in Fig. 95 (1). For certain values of the forces acting on it, the frame would be in equilibrium, but it would collapse if the forces were changed.

A *perfect or firm frame* is one which has a sufficient number of bars—and no more—to keep it in equilibrium for all systems of loading. Such a frame is shown at (2) in the figure.

A *redundant or over-firm frame* is one which has more bars than are necessary to keep it in equilibrium for all systems of loading. Such a frame is shown at (3) in the figure.

Objections to Deficient and Redundant Frames. If a deficient frame is actually pin-jointed, it is in unstable equilibrium; if its joints are riveted, then its stability depends on the stiffness of the joints and its members are subjected to bending stresses which it is the object of the framework to avoid.

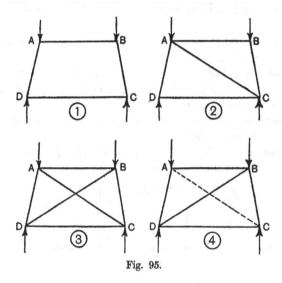

Fig. 95.

Redundant frames have the following disadvantages:

(1) Any stress in one member caused by bad fitting or change of temperature causes stresses in all the other members.

(2) The stresses in the members cannot be calculated by any simple mathematical or graphical process.

Such frames are sometimes called " statically indeterminate."

Semi-member or Counterbraced Frames. Some frames, which have the appearance of redundant frames, act as perfect frames and may be treated as such. Fig. 95 (4) shows such a frame. There are two diagonal bars *BD* and *AC*, but each can act in tension only, so that if the loading is such as would tend to put one of the diagonals, say *AC*, in compression, such diagonal would go out of action and the frame would act as if *BD* were the only diagonal.

The diagonals *AC* and *BD* are called *semi-members* or *counterbraces* and are commonly used in practice, especially in the centre

panels of railway bridge trusses in which the crossing of the
load causes a reversal of the stress in the diagonals.

**Relation between Bars and Nodes in a Perfect or Firm
Frame.** Consider a firm frame such as is shown at (2) in Fig. 95.
The first bar DC has 2 nodes.

It requires two more bars AD and AC to produce the next
node A, and so on.

Therefore, if there are n nodes, 2 of them go to the first bar
and the remaining $(n-2)$ require $2(n-2)$ bars.

∴ Total number of bars $= 2(n-2) \times 1 = 2n-3$. Therefore, in a perfect or firm frame the number of bars is equal to
twice the number of nodes minus 3.

If the number of bars is more than this, the frame is redundant;
if less, the frame is deficient.

The student should test this relation with the framed structures
shown in the following figures.

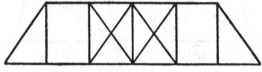

Fig. 96.

The converse of the above statement does not hold. The
number of bars might be $= 2n-3$, and yet the frame might
not be perfect.

Fig. 96 gives an example of this. In this case the number
of nodes is 12 and the number of bars 21, so that this fulfils
the above condition, although it is not a perfect frame.

Ties and Struts. If the force in a member of a structure
is a pull, the member is called a *tie*; if the force is a thrust, the
member is called a *strut*. The force in a bar transmitted from
a certain node will be a pull when the arrow-head points away
from the node, and a thrust when it points towards it.

It is desirable to distinguish between the ties and the struts
in the drawing of a framed structure. This can be done in any
of the following ways:

(1)   By drawing the struts in thicker lines than the ties.

(2)   By drawing a short single line across the ties and a
double line across the struts, e.g., | and ‖.

(3) By indicating the struts with a *plus* sign and the ties with a *minus* sign.

Loading of Framed Structures. Framed structures must always be taken as loaded at the nodes only. If a given bar is loaded between the nodes, then it acts as a beam and distributes to the nodes at each end the reaction of the beam.

Curved Members in Framed Structures. In some cases the members or bars of a framework are curved. For obtaining the forces in the bars (not really the *stresses*, although this term is most often used), we replace the curved bars by straight ones; but it must be carefully remembered that such bars will actually be subjected also to bending stresses which must be allowed for in design.

Determination of force in a Framed Structure. Since a pin-joint is provided at each node, the forces in the bars meeting in such a node and the external forces must be in equilibrium.

Before the forces in a frame can be determined, the whole of the external forces, including the reactions at the supports, must be known.

Example of simple Roof-Truss. Take for example the simple roof-truss shown in Fig. 97. The external forces acting are the weight $W = 2000$ lbs. at C and the reactions R_A, R_B at A and B. We assume that the truss simply rests on its supports so that these reactions are vertical. Since the frame looks the same whether viewed from the front or the back, or "from considerations of symmetry" as this is usually expressed, the reactions must be equal to each other, and for the equilibrium of the structure as a whole, their sum must be 2000 lbs.

$$\therefore R_A = R_B = 1000 \text{ lbs.}$$

Now consider the node C. There are three forces acting there, viz., W vertically downwards and forces f_{AC}, f_{BC} in AC and CB; and these forces must be represented by a triangle.

$\therefore$ draw 1, 2 to represent $W = 2000$ lbs. to a convenient scale and draw 1, 3 parallel to CA and 2, 3 parallel to CB to intersect at 3. Then 2, $3 = f_{BC}$ and 3, $1 = f_{AC}$. On scaling these off, we find $f_{BC} = f_{AC} = 2240$ lbs. (Thrust, because the arrow-head points towards the node).

Next consider the node A. The forces acting are the reaction R_A vertically upwards and f_{CA}, f_{BA} in CA and BA.

∴ draw 4, 1 to represent $R_A = 1000$ lbs. and draw 4, 3 parallel to BA and 1, 3 parallel to CA to intersect in 3. Scaling these off we find $f_{BA} = 3, 4 = 1000$ lbs. (Pull, because the arrowhead points away from the node), $f_{CA} = 2240$ lbs. (Thrust, as before).

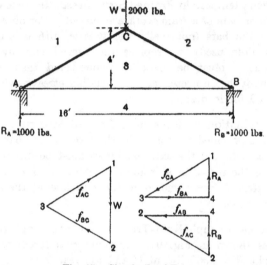

Fig. 97.   Simple Roof-Truss.

It will be noted that the force transmitted to any bar from the nodes at its two ends must be equal and opposite; otherwise that bar would not be in equilibrium.

The Reciprocal Figure Construction. The reciprocal figure construction is an extension, devised by Clerk Maxwell, of the method that we have just explained; the various vector polygons for each node are combined in one diagram. We will first explain the construction with reference to the roof-truss shown in Fig. 98. We will take the vertical loads on the nodes as equal, the reactions then being equal and vertical.

To commence the reciprocal figure set down lengths 1, 2; 2, 3; etc., on a vertical line to represent the forces, to some convenient scale, the reaction 4, 5 being equal to half the total load, and giving the point 5 as shown. At the left-hand end of the truss three lines meet, viz., 5, 1; 1, A; A, 5. On the reciprocal figure we require a corresponding triangle, so draw 1, a parallel to 1, A, and 5, a parallel to 5, A, their point of intersection determining the point a on the reciprocal figure. From a draw ab parallel to AB, and $2b$ parallel to $2B$, thus obtaining the point b; then bc parallel to BC, and $5c$ parallel to $5C$, thus obtaining the point c, and so on.

Frame diagram

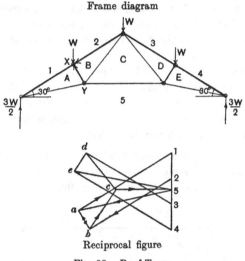

Reciprocal figure

Fig. 98.   Roof-Truss.

To serve as a check on the accuracy of the drawing, the line joining the last point e on the reciprocal figure to the point 5 should be parallel to the bar $E5$ of the frame.

Then the lengths of the lines of the reciprocal figure give— to the scale to which the loads were set down—the forces in the corresponding bars of the frame.

To distinguish between Ties and Struts. To ascertain which members of a framework are ties and which are struts, the following method is adopted and can be applied for all systems of loading.

Consider any one of the nodes of the truss at which the direction of one force is known, say the node X. Corresponding to this node we have the polygon $12ba1$ on the reciprocal figure. The direction of the force 12 is known to be vertically downward, so continue the arrow-heads in this direction round the polygon 12*ba*1. Now transfer the direction of these arrow-heads to the corresponding bars close to the given node. Then if the arrow-head on a given bar points towards the node, the bar is a strut; and if it points away, the bar is a tie. In this way it is seen that the bars 1*A*, *AB*, and *B*2 are all struts.

Now consider the node *Y*. Corresponding to this we have the polygon 5*abc*5. Since *AB* is a strut, the arrow-head at the node *Y* points towards the node, and so the arrow-heads go round the polygon in the direction *ab*, *bc*, *c*5, 5*a*, as shown. Transferring these arrow-heads to the frame diagram, we see that the bars *BC*, *C*5, and 5*A* are all ties.

With practice one can tell by inspection in most cases whether a given bar is a strut or a tie by the following rule: If, on imagining the given bar cut through, the forces would tend to increase its length, such bar is a tie; if the forces would tend to decrease its length, the bar is a strut.

Example of Warren Girder with Uneven Loading. As a further example of reciprocal figures, take the example of the Warren girder loaded as shown in Fig. 99.

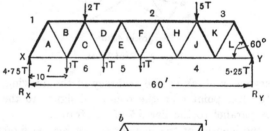

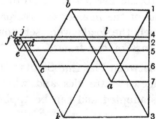

Fig. 99. Warren Girder.

Unless it is definitely stated to the contrary we can always take it that in framed girders the "panels" or "bays" as they are

called are of equal length, i.e. in this case the span is divided into
six equal parts.

We must first find the reactions before we can proceed with
the reciprocal figure. Taking moments about X we have

$$R_Y \times 60 = 1 \times 10 + 1 \times 20 + 1 \times 30 + 2 \times 15 + 5 \times 45 = 315;$$
$$\therefore R_Y = \tfrac{315}{60} = 5 \cdot 25 \text{ tons};$$
$$\therefore R_X = 10 - 5 \cdot 25 = 4 \cdot 75 \text{ tons}.$$

Choosing a suitable force scale, we set down the vertical forces
in order, i.e. first set down 1, 2 and 2, 3 to represent 2 and
5 tons respectively; next set up 3, 4 to represent the reaction
of 5·25 tons; and then set down 4, 5; 5, 6; and 6, 7 to represent
1 ton each, the length 7, 1 checking back to give the reaction
4·75 tons. We now proceed as before, drawing 1a parallel
to 1A, and 7a parallel to 7A; then ab and 1b parallel respectively
to AB and 1B, and so on, the reciprocal figure coming as shown,
and l3 coming parallel to L3, and thus serving as a check on the
drawing.

In cases of complicated frames where some difficulty is
experienced of getting the last line to check, it is well to start
the reciprocal figure from each end of the frame, the errors
being in this way minimised.

Figs. 100, 101 show the reciprocal figures for two other common
forms; the student should work these as an exercise; with a
little practice it will be found that these diagrams can be drawn
without difficulty and they have the great advantage that the
closing line forms a check on the work.

There need be no hard and fast rule as to the end from which
we commence the figure. In Fig. 101, for instance, we have
commenced from the right-hand side whereas in the other cases
we have commenced from the left-hand side. When two points
on a reciprocal diagram coincide, and the line joining them
corresponds to a bar in the frame, the force in the corresponding
bar is zero. Thus in Fig. 101 the stress in FF' is zero and so
we have drawn the bar in dotted lines; the other bar shown
dotted being a semi-member or counterbrace (see p. 175). The
bar FF' is not really redundant because if the loading were
altered, there would be a stress in it.

Stresses in Framed Structures by Moments. The stresses
in framed structures can also be found by the following method,

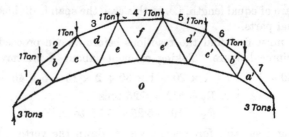

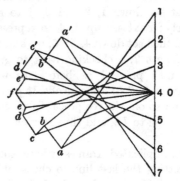

Fig. 100. Crescent Roof-Truss.

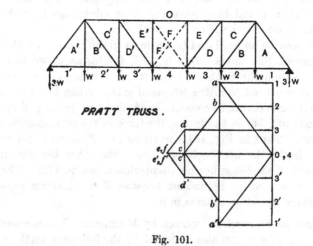

PRATT TRUSS.

Fig. 101.

which is called the method of Moments or Sections, or sometimes
Ritter's method. The method consists in imagining one bar
to be cut through and in finding the point about which the
structure tends to collapse. Consider for example the simple
roof-truss shown in Fig. 102 and suppose that the bar AB is

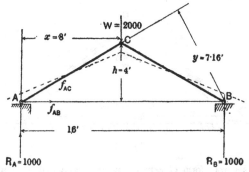

Fig. 102. The method of Moments.

cut through; the ends A and B will then move outwards somewhat as indicated in dotted lines and the bars AC, BC will
pivot about the point C. Therefore the force f_{AB} in AB must
be the force which prevents this pivoting movement and its
moment about the point C must be equal to the moment of
either reaction about C.

$$\therefore f_{AB} \times h = R_A \times x,$$

i.e. $$f_{AB} = \frac{1000 \times 8}{4} = 2000 \text{ lbs.}$$

In using this method it is best to regard one side of the
structure as remaining fixed and the other side as moving under
the action of the forces upon it; in the present example therefore
we regard BC as fixed and AC as moving upward, pivoting
about C.

Now let us find the force f_{AC} in AC. If AC were cut through
the weight W would fall down and the bar CB would pivot
about the point B. The only force tending to cause this collapse
is W whose moment about $B = 2000 \times 8$,

$$\therefore f_{AC} \times y = 2000 \times 8,$$
$$f_{AC} = \frac{2000 \times 8}{y} = \frac{2000 \times 8}{7 \cdot 16}$$
$$= 2240 \text{ lbs. nearly.}$$

Experiment upon Model Roof-Truss. A simple experimental form of roof-truss similar to that which we have already considered is shown in Fig. 103. The struts AC, CB (called *rafters*) are each formed of two parts sliding one within the other and connected by compression spring balances which will measure the forces in these bars. The tie-bar AB consists of two pieces of wire or string connected by an ordinary spring balance which will measure the pull in it.

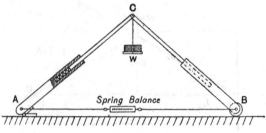

Fig. 103.

A more accurate form of experiment consists in making a model frame in iron and measuring the small changes in length of the various members by means of a very sensitive instrument called the *extensometer* (see p. 147), from the readings of which the forces acting in the various bars can be readily calculated.

Forces in Tripods and Shear Legs. In these cases we proceed as follows:

Draw the structure in plan and elevation, and let W be the load at A, Fig. 104, AB being the back leg and AD, AE the fore

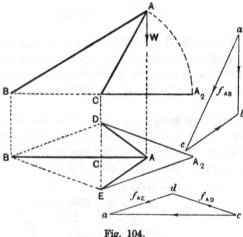

Fig. 104.

legs. Resolve W down AB and *down the plane* of the other two legs, i.e., set out ab equal to W and draw bc parallel to AB, and ac parallel to AC, then bc represents the force in AB. Now swing the shear legs down horizontally in order to get A_2DE, the true shape of the triangle ADE, then setting out ac horizontally and drawing ad and cd parallel to EA_2, DA_2 respectively, we get the forces in the fore legs.

SUMMARY OF CHAPTER XI.

Frames may be deficient, firm or redundant, but only in the case of firm frames can the forces in the members be found by simple methods.

Before the forces in the various members of a frame can be calculated, all the forces acting upon it, including the reactions, must be known.

At each node, the forces in the members meeting there and the external forces acting there must be in equilibrium and so must form a closed polygon.

Members subjected to pulling forces are called *ties* and those subjected to thrusting forces are called *struts*. The reciprocal figure construction forms an automatic check upon its accuracy because the closing line must be parallel to the corresponding bar in the frame.

The method of moments enables the force in any particular bar to be calculated and by applying the method to a convenient bar, preferably near the middle of the structure, we get a useful check upon the accuracy of the graphical construction.

EXERCISES. XI.

1. Find the forces in the members of the derrick crane shown in Fig. XI *a*.

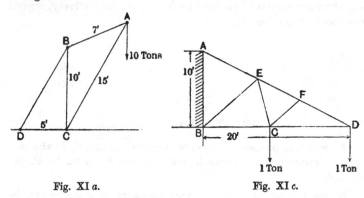

Fig. XI *a*.   Fig. XI *c*.

2. Find the forces in the members of the framed structure shown in Fig. XI *b* and check your result for *BB'* by moments.

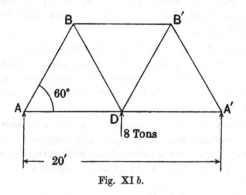

Fig. XI *b*.

3. A Warren girder of length 100 feet is divided into 5 bays on the lower flange, the length of the inclined braces being 20 feet. If loads of 30 tons are carried at the lower nodes 20 and 40 feet from one end find the forces in the members.

4. Find the forces in the members of the cantilever truss shown in Fig. XI *c*, which projects from a wall *AB*. [Note *BC = CD* and *AE = EF = FD*.]

5. Draw the reciprocal figure for the roof-truss shown in Fig. XI d given that $AB = BC = CB' = B'A'$; $AD = A'D' = CD = CD'$ and scale off the force in DD'.

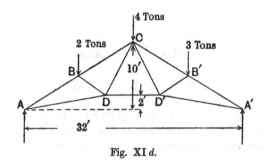

Fig. XI d.

6. A load of 7 tons is suspended from a tripod the legs of which are of equal length and inclined at 60° to the horizontal. Find the thrust on each leg.

7. A pair of shear legs (Fig. XI e) make an angle of 20° with each other and their plane makes an angle of 60° with the horizontal. The back stay is at an angle of 30° to the horizontal. Find the force in each leg and in the stay when supporting a load of 10 tons.

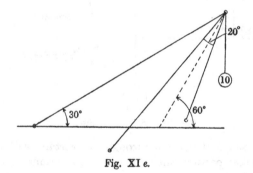

Fig. XI e.

CHAPTER XII

BEAMS AND GIRDERS

WE shall get a good preliminary idea of the stresses occurring in beams by considering a model devised by Prof. Perry. Suppose that a beam fixed at one end carries a weight W (Fig. 105) at the

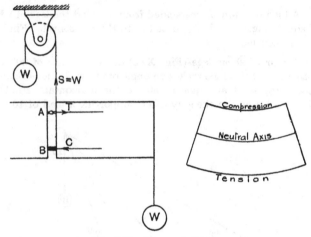

Fig. 105. Stresses in Beams.

other end, and that it is cut through at a certain section. Then the right-hand portion can be kept in equilibrium by attaching a rope to the top and passing over a pulley, a weight W being attached to the other end of the rope, and by placing a block B at the lower portion of the section and a chain A at the upper portion. Then the pull in the rope overcomes what is called the *shearing force*; and the block B carries a compressive force C, and the chain A carries a tensile force T. Since these are the only horizontal forces, they must be equal and opposite, and thus form a *couple*. Then the moment of this couple must be equal and opposite to the couple due to the loading, which is called the bending moment.

In the actual beam, owing to the deflection which takes place, the material on one side of the beam will be stretched, and the material on the other side will be compressed, so that at some point between the two sides the material will not be strained at all, and the axis in the section of the beam at which there is no strain is called the *neutral axis*. To tell whether the top or the bottom is in tension we consider the deflected form which the beam will take up and note that the tension edge is always on the outside of the bend whereas the compression edge is on the inside.

Shearing Force and Bending Moment. The actual calculation of the stresses in a beam is beyond our present scope but such calculations depend upon the quantities called *Shearing Force* and *Bending Moment* which are quite simple to understand and with which we will now deal at some length.

Definitions. The *shearing force* at any point along the span of a beam is the algebraic sum of all the perpendicular forces acting on the portion of the beam to the right or to the left of that point.

The *bending moment* at any point along the span of a beam is the algebraic sum of the moments about that point of all the forces acting on the portion of the beam to the right or to the left of that point.

As the beam is in equilibrium under the forces acting on it, at any point the algebraic sum of the forces, and of the moments of the forces about the point, acting *on both* sides must be nothing; so that we shall get the same numerical values for the shearing force and bending moment from whichever side we consider them, but they will be opposite in sign. We will, wherever possible, always consider the shearing force and bending moment of the forces to the right of the section, and we will take an *upward* shearing force and a *clockwise* bending moment as positive, the downward and anti-clockwise being taken as negative.

Bending Moment and Shearing Force Diagrams. If the bending moment and shearing force at every point of the span be plotted against the span and the points thus obtained be joined up, we shall get two diagrams called the Bending Moment (B.M.) and Shear diagrams, and from these diagrams the values of these quantities can be read off at any point of the span. We will examine the forms of these diagrams for various kinds of loading and for various ways of supporting the beam, and will first consider beams with fixed loads. We will use M_P and S_P to represent respectively the bending moment and shearing force at a point P.

We will restrict our consideration to loads which are fixed in position as opposed to those which may roll from one position to another.

A.   Cantilevers, i.e. beams fixed at one end and free at the other, the loads being all at right angles to the length of the beam.

CASE 1.   CANTILEVER WITH ONE ISOLATED LOAD. Let a cantilever, fixed at the end B, Fig. 106, carry an isolated load W at the point A, at distance l from B. Consider any point P at distance x from A.

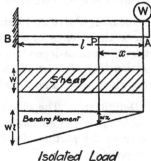

Then we have

$$S_P = W.$$

This is constant throughout the span.

Isolated Load

Fig. 106.   Cantilevers.

∴ Shear diagram is a rectangle of height W.

Again $M_P = W \times x$.

This is proportional to x.

∴ B.M. diagram is a triangle whose maximum ordinate is Wl, this being the bending moment at the point B.

CASE 2.   CANTILEVER WITH UNIFORM LOAD. Let a uniformly distributed load of p tons per foot run be carried by a cantilever AB of span l, Fig. 107. Consider a point P at distance x from the free end A.   Then

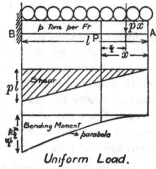

Uniform Load.

Fig. 107.   Cantilevers.

$$S_P = \text{load on } AP$$

$$= px.$$

This is proportional to x, and therefore the shear diagram is a triangle, the maximum shear occurring at the end B, and being equal to pl or W, if W is the total load on the cantilever.

M_P = Moment of load px about P

$$= px \times \frac{x}{2}$$

$$= \frac{px^2}{2}.$$

This is proportional to x^2, and therefore the B.M. diagram will be a parabola with vertex at A. The maximum B.M. will be equal to $\frac{pl^2}{2}$ or $\frac{Wl}{2}$ and occurs at B.

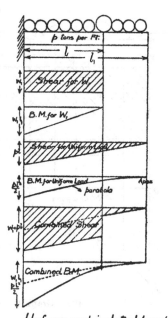

Uniform and isolated Loads

Fig. 108.   Combined loading on Cantilevers.

CASE 3. CANTILEVER WITH ISOLATED LOAD AND UNIFORM LOAD. Since the B.M. and shear at any point are defined as the sum of the moments and the forces to the right of that point, it follows that the B.M. and shear diagrams for a number of loads can be obtained by adding together the diagrams for the separate loads.

In adding together two diagrams we first draw the separate diagrams and then make diagrams whose ordinates at each point are the sums of the ordinates of the separate diagrams at the same point.

CASE 4. CANTILEVER WITH IRREGULAR LOAD SYSTEM.— GRAPHICAL METHOD. Suppose a number of loads 0, 1, 1, 2, and so on, Fig. 109, act on a cantilever. To obtain the shear and B.M. diagrams set down 0, 1, 1, 2, 2, 3, etc., down a vector line 0, 5 to represent the forces to some convenient scale, and take a pole P at some convenient distance p from the vector line 0, 5 and join P to each of the points 0 to 5 on the vector line.

Now across the lines of the forces draw ag parallel to $P0$;

across space 1 draw *ab* parallel to *P*1; across space 2 draw *bc* parallel to *P*2, and so on until the point *f* is reached.

Then *abcdefg* is the B.M. diagram.

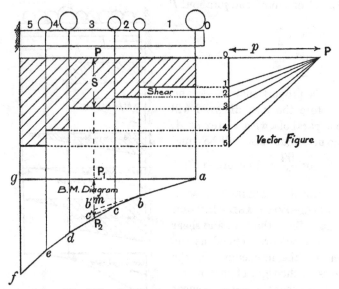

Fig. 109. Graphical method for Cantilevers.

To obtain the shear diagram, project the points 0–5 on the vector line across their corresponding spaces, the line through the point 0 being drawn right across the span, the stepped figure thus obtained being the shear diagram.

PROOF. Consider any point *P* along the span, and produce *ab* and *bc* to cut the corresponding ordinate $P_1 P_2$ of the link polygon at b' and c' respectively.

Now consider the triangles $aP_1 b'$ and $P01$.

They are similar, and as the bases of similar triangles are proportional to their heights, we have

$$\frac{P_1 b'}{0,1} = \frac{aP_1}{p};$$

$$\therefore p \times P_1 b' = 0, 1 \times aP_1.$$

But $0, 1 \times aP_1 =$ Moment of force $0, 1$ about P.

$$\therefore p \times P_1 b' = \text{Moment of force } 0, 1 \text{ about } P.$$

Similarly it follows that

$$p \times b'c' = \text{Moment of force 1, 2 about } P,$$

and $\qquad p \times c'P_2 = \text{Moment of force 2, 3 about } P.$

$\therefore$ we see that $p \times P_1P_2 = p \, (P_1b' + b'c' + c'P_2)$

$\qquad\qquad = \text{Moment of all forces to left of } P \text{ about } P$

$\qquad\qquad = M_P.$

$\therefore$ since p is a constant quantity, it follows that the ordinates of the link polygon represent the bending moments at the corresponding points of the beam.

Now consider the shear S at P. The total force to the left of P is $0, 1 + 1, 2 + 2, 3 = 0, 3$, and this is obviously the value given on the shear diagram.

SCALES. In all graphical constructions it is extremely important to state clearly the scales to which the various quantities are plotted, and to see that such scales are convenient for reading off.

Let the space scale be

$$1 \text{ in.} = x \text{ feet}$$

and the load scale on the vector line

$$1 \text{ in.} = y \text{ tons,}$$

and let the polar distance be p actual inches.

Then the scale to which the bending moments can be read off is $1 \text{ in.} = p \times x \times y$ tons-ft.

p should thus be chosen so as to make this a convenient round number.

To take a numerical example, suppose the space scale is $1 \text{ in.} = 4 \text{ ft.}$ and the load scale is $1 \text{ in.} = 2 \text{ tons}$, then if p is taken as $2\frac{1}{2}$ ins. the B.M. scale will be $1 \text{ in.} = 4 \times 2 \times 2\frac{1}{2} = 20$ tons-ft.

If p had been taken 2 ins. the B.M. scale would be $1 \text{ in.} = 16$ tons-ft. which would not be nearly such a convenient scale.

B. Simply Supported Beams, i.e. beams simply resting on two supports, the loading all being at right angles to the length of the beam. Unless it is definitely stated to the contrary, we will always take it that the supports are at the ends of the beam.

In simply supported beams the forces acting are the loads and the reactions at the supports, the sum of the reactions being equal to the total load, and their values being obtained by means of moments as explained in Chapter II. As the ends are freely supported, there can be no bending moment at either end.

We will now consider the following standard cases:

CASE 1. ISOLATED LOAD IN ANY POSITION. Let a load W be supported at a point C, Fig. 110, on a beam AB of span l, the distances of the point C from B and A being b and a respectively.

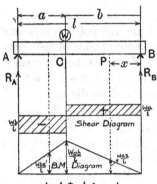

To obtain the reaction R_B at B take moments round A.

Then
$$R_B \times l = W \times a,$$
$$R_B = \frac{W \times a}{l}.$$

Similarly
$$R_A = \frac{W \times b}{l}.$$

Now consider a point P between B and C.

Isolated Load

Fig. 110.

$$S_P = R_B = \frac{+ Wa}{l};$$

∴ between B and C the shear diagram is represented by a rectangle of height
$$= \frac{Wa}{l}.$$

Now take a point between C and A.

Shear $= R_B - W$
$$= \frac{Wa}{l} - W = W \left(\frac{a - l}{l}\right) = \frac{- Wb}{l} = - R_A;$$

∴ Shear between C and A is represented by a rectangle of height
$$= \frac{- Wb}{l}.$$

In the case of the cantilever there was no need to distinguish between positive and negative shear because there was no change in direction of the shear; but in the present case there is a change in direction, and so we will use the rule given on p. 189.

Now considering the bending moment,

$$M_P = R_B \times x = \frac{W \cdot a \cdot x}{l}.$$

This is proportional to x, and therefore the B.M. diagram between B and C will be a triangle, the B.M. at C being equal to

$$\frac{Wab}{l} = \frac{Wa\,(l-a)}{l}.$$

If P were between C and A and at distance x' from A we should have

$$\begin{aligned}
M_P &= R_B\,(l-x') - W\,(l-x'-b) \\
&= R_B \cdot l - R_B\,x' - Wl + Wx' + Wb \\
&= x'\,(W - R_B) + Wb - l\,(W - R_B) \\
&= R_A \cdot x' + Wb - lR_A \\
&= \frac{Wbx'}{l} + Wb - Wb \\
&= \frac{Wbx'}{l}.
\end{aligned}$$

This is proportional to x', and therefore the B.M. diagram between A and C is also a triangle, the whole diagram then being as shown in the figure.

CASE 2. ISOLATED LOAD AT CENTRE. This is a special case of the preceding one, in which $a = b = \frac{l}{2}$.

Each reaction is now equal to $\frac{W}{2}$ and the maximum B.M.

$$= \frac{W \times \frac{l}{2} \times \frac{l}{2}}{l} = \frac{Wl}{4}.$$

CASE 3. UNIFORM LOAD OVER WHOLE SPAN. Let a uniform load of p tons per ft. run cover the whole span AB, Fig. 111, and consider a point C at distance x from B.

In this case the two reactions will, from symmetry, be equal, and each have the value $\frac{pl}{2}$ or $\frac{W}{2}$.

Then $\qquad S_C = R_B - px = p\left(\frac{l}{2} - x\right).$

This is a linear relation, therefore the shear diagram will be a triangle as shown, having values $\pm \dfrac{pl}{2}$ at the ends and changing sign at the centre.

Now consider the bending moment.

$$M_C = R_B \times x - px \times \frac{x}{2}$$

$$= \frac{plx}{2} - \frac{px^2}{2} = \frac{p}{2}(lx - x^2).$$

This depends on x^2, and therefore the B.M. diagram will be a parabola.

The maximum B.M. will occur at the centre, i.e. when $x = \dfrac{l}{2}$.

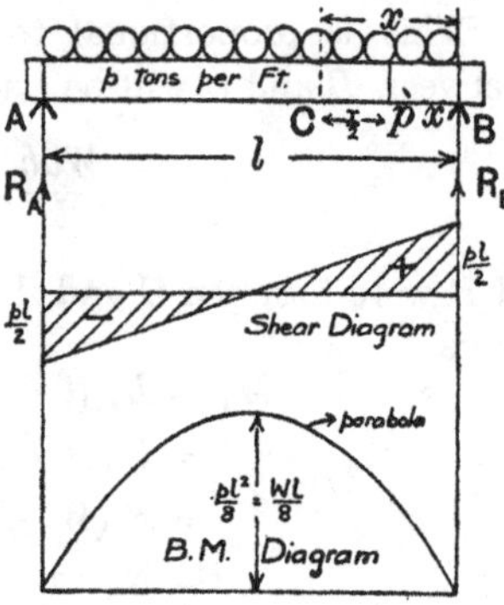

Uniform load over whole span

Fig. 111.

Then maximum B.M.

$$= \frac{p}{2}\left[\left(\frac{l \cdot l}{2}\right) - \left(\frac{l}{2}\right)^2\right] = \frac{p}{2}\left(\frac{l^2}{2} - \frac{l^2}{4}\right)$$

$$= \frac{p}{2} \times \frac{l^2}{4} = \frac{pl^2}{8} \text{ or } \frac{Wl}{8}.$$

CASE 4. IRREGULAR LOAD.—GRAPHICAL CONSTRUCTION. Let a number of loads W_1, W_2, W_3, and W_4 be placed anywhere along a span AB, Fig. 112. Number the spaces between the loads and set down 0, 1; 1, 2; 2, 3; 3, 4 as a vertical vector line to represent the loads to some convenient scale, and in *any position* take a point P at a suitable polar distance p from the vector line, and join $P0$, $P1$, $P2$, etc.

Across space 0 then draw ab parallel to $P0$; across space 1 draw bc parallel to $P1$ and so on until ef is reached, this being parallel to $P4$.

Join af, then the figure a, b, c, d, e, f, a will give the B.M. diagram for the given load system.

Now draw Px parallel to af, the closing link of the link polygon; then on the vector line, $4x = R_B$ and $x0 = R_A$.

To draw the shear diagram, draw a horizontal line through x right across the span: this gives the base line for shear. Now project the point 0 horizontally across space 0; project point

1 across space 1 and so on, the stepped diagram thus obtained being the shear diagram.

If the first and last links are produced to meet at Y, then as we proved on p. 28, the resultant load acts through Y.

PROOF. By reasoning by similar triangles as for the cantilever we can prove that $4x = R_B$ and $x0 = R_A$.

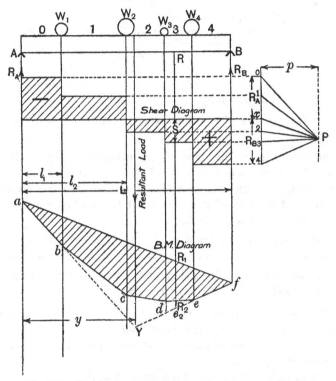

Fig. 112. Graphical construction for simply supported beam.

Now consider any point R along the span.

$$S_R = R_B - W_4$$
$$= 4x - 3, 4 = 3x,$$

but the ordinate S of the shear diagram is equal to $3x$, and therefore the stepped figure gives the correct shearing force at any point.

Let the vertical through R cut the B.M. diagram in R_1R_2 and fe produced in e_2.

Then by exactly similar reasoning as before

$$R_1e_2 = \frac{\text{Moment of } R_B \text{ about } R}{p},$$

$$R_2e_2 = \frac{\text{Moment of } W_4 \text{ about } R}{p},$$

$$\therefore R_1R_2 = R_1e_2 - R_2e_2$$

$$= \frac{\text{Moment of } R_B - \text{Moment of } W_4 \text{ about } R}{p}$$

$$= \frac{M_R}{p};$$

$$\therefore M_R = p \times R_1R_2.$$

$\therefore$ the ordinate of the B.M. diagram represents the B.M. at any point.

SCALES. As in the case of the cantilever (page 193), if $1'' = x$ feet is the space scale and $1'' = y$ tons is the force scale, and if the polar distance is p actual inches, then the vertical ordinates of the B.M. diagram represent the bending moment to a scale $1'' = p \times x \times y$ tons ft.

NOTE.—In this construction the bending moment R_1R_2 is measured *vertically* and not at right angles to the closing line af.

The above construction is a special case of the link and vector polygon construction described on p. 28.

Numerical Examples. (1) *A freely supported beam of 20 ft. span carries a uniformly distributed load of 5 tons, and isolated loads of 3 and 2 tons, at distances respectively of 4 and 5 ft. from the ends (see Fig. 113).*

We have first to get the reactions R_A and R_B.

Take moments round B.

$$R_A \times 20 = 5 \times 10 + 3 \times 16 + 2 \times 5$$
$$= 50 + 48 + 10 = 108,$$

$$\therefore R_A = \frac{108}{20} = 5\cdot4 \text{ tons,}$$

$$\therefore R_B = 10 - 5\cdot4 = 4\cdot6 \text{ tons.}$$

The shear diagram is then as shown in the figure, the amounts

of the steps being equal to the isolated loads. The point at
which the shear is nothing is found as follows:

Let it be at distance x from B. Then

$$S_x = 0 = R_B - 2 - p \cdot x$$

$$= 4 \cdot 6 - 2 - \frac{5x}{20}$$

$$= 2 \cdot 6 - \frac{x}{4};$$

$$\therefore \frac{x}{4} = 2 \cdot 6,$$

$$x = 10 \cdot 4 \text{ feet.}$$

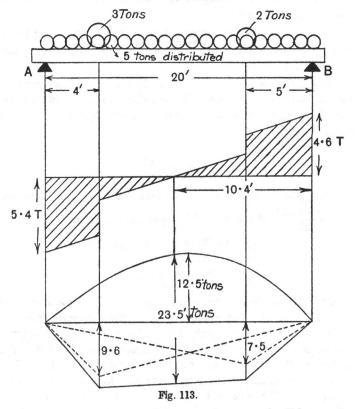

Fig. 113.

The B.M. at this point will be a maximum, and will be equal to

$$M_x = R_B \times 10 \cdot 4 - 2 (10 \cdot 4 - 5) - \frac{1}{4} \cdot \frac{10 \cdot 4^2}{2}$$

$$= 47 \cdot 84 - 10 \cdot 8 - 13 \cdot 52$$

$$= 23 \cdot 52 \text{ tons-ft.}$$

The B.M. diagram will consist of a parabola for the uniformly distributed load, the maximum ordinate of which is equal to

$$\frac{5 \times 20}{8} = 12\cdot5 \text{ tons-ft.}$$

The B.M. diagram for each of the isolated loads will be a triangle, the respective heights being

$$\frac{3 \times 4 \times 16}{20} = 9\cdot6 \text{ tons-ft. and } \frac{2 \times 5 \times 15}{20} = 7\cdot5 \text{ tons-ft.}$$

Combining these three figures we get the B.M. diagram shown on the figure, and on scaling off the maximum ordinate it will be found to be 23·5 tons-ft.

NOTE.—In all constructions where diagrams are going to be added together, such diagrams must of course be drawn to the same scale.

(2) *A certain joist used as a cantilever weighs* 18 *lbs. per foot, and the maximum B.M. which it can safely carry is* 63·56 *tons-ins. Find how long the span may be for the cantilever to be able to sustain safely its own weight.*

We have for a cantilever

$$\text{Max. B.M.} = \frac{pl^2}{2};$$

Now $p = 18$ lbs. per ft.; $\therefore$ max. B.M. = 63·56 tons-ins.

$$= \frac{63\cdot56 \times 2240}{12} \text{ lbs.-ft.; } \therefore \text{ if } l \text{ is span in feet we have}$$

$$l^2 = \frac{2 \times 63\cdot56 \times 2240}{12 \times 18} = 1318.$$

$$l = \sqrt{1318} = \underline{36\cdot3 \text{ feet.}}$$

(3) *A beam* 20 *ft. span carries loads of* $\frac{1}{2}$, $\frac{1}{4}$, 1 *and* 2 *tons, as shown on Fig.* 15. *Determine graphically the maximum B.M.*

Draw the B.M. curve by the link and vector polygon construction as shown in Fig. 22. Take the space scale $1'' = 4$ ft.; the load scale $1'' = 2$ tons; and the polar distance $1\frac{1}{4}$ inches. The maximum ordinate of the B.M. curve will then be found to be 1·09 inches. The scale of this will be $1'' = 1\frac{1}{4} \times 4 \times 2 = 10$ tons-ft

$$\therefore \underline{\text{Maximum B.M.} = 10\cdot9 \text{ tons-ft.}}$$

SUMMARY OF CHAPTER XII.

Neutral axis of a beam is the line in the cross section which receives no stress or strain.

The stresses in a beam are tensile on one side of the neutral axis and compressive on the other, the resultant forming a couple whose moment must be equal to the bending moment.

The *shearing force* at any point is the algebraic sum of all the perpendicular forces acting on the portion of the beam to the right *or* left of that point.

The *bending moment* at any point is the algebraic sum of the moments about that point of all the forces acting to the right *or* left of that point.

Bending moment and shear diagrams can be drawn for standard methods of loading and fixing the ends of the beam; for two or more loadings occurring together, the separate diagrams are added together.

For a simply supported beam of span l with an isolated load W at distance a from one end

$$\text{Max. B.M.} = \frac{Wa\,(l-a)}{l}$$

$$\text{For uniform load Max. B.M.} = \frac{Wl}{8}.$$

Shear and bending moment diagrams for any loading may be drawn by means of the link and vector polygon construction.

EXERCISES. XII.

1. A beam AB 15 ft. long is fixed at A and free at the other end. A weight of 80 lbs. is placed at the end B, and a weight of 100 lbs. in the middle of the beam. Find the B.M. and S.F. at distances of 3, 6 and 9 ft. from the fixed end.

2. A beam 150 ft. span is uniformly loaded with 2 tons per ft. run. Calculate the B.M. and S.F. at every 10 ft. of its span and draw the curves of B.M. and S.F.

3. A girder supported at both ends is 50 ft. span and carries a uniformly distributed load of 200 lbs. per ft. run. Find the B.M. and S.F. at the centre and at points 15 ft. from the ends.

4. A cantilever is 35 ft. long and is uniformly loaded with 150 lbs. per ft. run. Find the B.M. and S.F. at the fixed end and at points 25 ft. and 15 ft. from the fixed end.

5. If in the last question an additional load of 750 lbs. is placed in the centre of the beam find the magnitude of the B.M. and S.F. at the points mentioned.

6. A beam 45 ft. span supported at the ends carries a weight of 6 tons 15 ft. from one end. Find the B.M. and S.F. at the centre and also at 5 ft. from each end. Draw a diagram of B.M. and S.F. to scale.

7. A beam is loaded as shown (see Fig. XII a). Find the reactions at the supports, also the B.M. and S.F. at each quarter span, C, D and E being the points at quarter distance each along the beam.

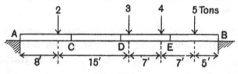

Fig. XII a.

8. A truck weighing 10 tons is carried on four wheels. The distance apart of the centres of rails is 5 ft. and of the axle boxes is 6 ft. 4 ins. Find the B.M. and S.F. at points on the axle 4 ins. apart. Draw diagrams.

9. A uniform beam, weighing 180 lbs. per ft. run, 20 ft. long is supported at the ends. It carries also a load of 60 lbs. in the centre. Find the B.M. and S.F. at points 5 ft. and 8 ft. from one end.

10. Find the B.M. and S.F. at the centre of a beam 50 ft. long weighing 500 lbs. per ft. run and carrying a load of 10,000 lbs. 20 ft. from one end. Draw the B.M. and S.F. diagrams.

11. A plank placed across an opening 12 ft. wide is broken by the bending effect of 3 cwt. placed 3 ft. from one end. What is the greatest load which a man weighing 156 lbs. could safely carry across?

12. Find the B.M. at the centre of a girder 40 ft. long supported at the ends and uniformly loaded with ½ ton per ft. run; there being an additional load of 5 tons 8 ft. from one end.

Find also the S.F. at the centre.

13. A timber beam is 18 ft. between supports and is 12 ins. deep by 4 ins. broad. Draw curves of B.M. and S.F. produced by its own weight; giving numerical values at each quarter span. Weight of timber 48 lbs. per cubic ft.

CHAPTER XIII

CENTRE OF GRAVITY AND CENTROID

THE **centre of gravity** of a solid body or of a number of bodies is a point through which the resultant weight of the whole body or bodies may be considered to act. Every portion of a body is attracted towards the centre of the earth by a force called the weight of that portion, and in a body of reasonable size the

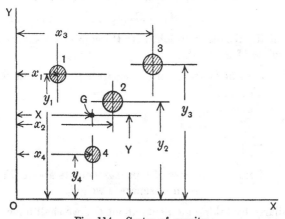

Fig. 114. Centre of gravity.

weights of all the parts will be parallel, so that the problem of finding the centre of gravity of a body resolves itself into that of finding the position of the resultant of a number of parallel forces, and this is solved, as we have already seen on p. 19, by the principle of moments.

Let 1, 2, 3, 4, etc., Fig. 114, represent a number of very small bodies in the same plane whose weights w_1, w_2, w_3, w_4, etc. act at right angles to the plane of the paper, the bodies being so small that the weight of each may be considered as acting through their centre. Suppose that G is their centre of gravity; then G is the point through which acts the resultant W of the parallel forces equal to the various weights. The magnitude W of this resultant will be the sum of the separate weights, i.e.

$$W = w_1 + w_2 + w_3 + w_4 + \dots.$$

For convenience this is written (as on p. 20)

$$W = \Sigma w_1 \dots\dots\dots\dots\dots\dots(1).$$

Now suppose that OX and OY are any two convenient lines which are not parallel; it is usually most convenient to take them at right angles to each other.

Then, remembering that the forces are acting at right angles to the plane of the paper we have, by taking moments about OY,

Total moment about OY

$$= M_Y = w_1x_1 + w_2x_2 + w_3x_3 + w_4x_4 + \dots,$$

i.e. $$M_Y = \Sigma w_1 x_1 \dots\dots\dots\dots\dots\dots(2).$$

But if the resultant weight acts through G we shall have by the principle of moments

$W \cdot X =$ Sum of moments of the separate forces about OY

$$= \Sigma w_1 x_1 ;$$

i.e. $$X = \frac{\Sigma w_1 x_1}{W}$$

$$= \frac{\Sigma w_1 x_1}{\Sigma w_1} \dots\dots\dots\dots\dots\dots\dots\dots(3)$$

$$= \frac{\text{Sum of moments of separate weights about } OY}{\text{Sum of separate weights}} .$$

Similarly by taking moments about OX we shall have

$$Y = \frac{\Sigma w_1 y_1}{\Sigma w_1} \dots\dots\dots\dots\dots\dots\dots\dots(4)$$

$$= \frac{\text{Sum of moments of separate weights about } OX}{\text{Sum of separate weights}} .$$

In this way we have fixed the distance of the centre of gravity from two given lines and so have found its exact position.

Application to continuous bodies. We can apply these principles to the determination of the centre of gravity of continuous bodies by imagining such bodies to be divided up into a very large number of very small parts, as indicated in Fig. 115, and regarding the weight of each separate part as acting through its centre. The greater the number of parts, the more accurate will our calculation be; as the number of parts becomes very great, how-ever, the calculation becomes very laborious and it is seldom made in this way but by means of a branch of mathematics called the *calculus*, which every engineering student should study if he wishes to understand easily the more advanced portions of mechanics.

Fig. 115.

The centre of gravity as the balance point. If a body balances about a point or a line then that point must be on the vertical line through the centre of gravity or the line must inter-sect that vertical line; moreover, if a body be freely suspended by a string or wire the wire or string must pass through the centre of gravity.

Take first the case of a body balanced about a point A, Fig. 116; there are only two forces acting upon the body, viz. the resultant weight W of the body and the upward pressure R at the support. Since the body is balanced, it must be in equilibrium under the action of these two forces and if two forces act upon a body and keep it in equilibrium they must be equal and opposite. If they were not, there would be a resultant moment about some points and the body would start turning.

The weight W acts vertically downwards, therefore the upward pressure R acts vertically upwards, so that the centre of gravity G must lie upon the vertical through the point of support A.

By exactly similar reasoning in the case of a body suspended by a string attached at a point B, the only forces acting are the weight W of the body and the tension T in the string; therefore the direction of the string must pass through the point G.

Now we have seen that the sum of the moments of a number of forces about any point is equal to the moment of the resultant about the same point, the resultant weight of a body passes through the centre of gravity and a force has zero moment about a point in its line of action. We therefore deduce the very important rule that *the sum of the moments about the centre of gravity of the weights of the separate portions of a body making up the whole body must be zero.* It is clear for instance from Fig. 116

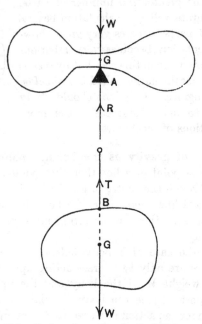

Fig. 116. Centre of gravity as the balance point.

that the moments of the portions to the left of G are anti-clockwise, while those to the right are clockwise and therefore of opposite sign.

Centre of gravity by inspection. The centre of gravity of a body which possesses a section of symmetry will always be in that section. By a section of symmetry is meant a section which will divide it into two exactly similar parts, which are "looking-glass pictures of each other."

If a body has two different sections of symmetry, the centre of gravity will always be on the intersection of the two sections.

Take for instance the cylinder represented in Fig. 117 which is assumed to be of the same material throughout. Three sections of symmetry are shown; one vertical, cutting the cylinder along the centre of its length: one at right angles to this, also vertical and cutting through the centre of the two ends: and the

third horizontal, also cutting through the centres of the two ends.
The centre of gravity G is at the intersection of the three sections.

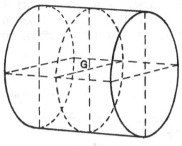

Fig. 117.

As a proof of the statement that the centre of gravity must
lie upon a section of symmetry, consider the body shown in
Fig. 117 a of which XX is a section of symmetry. For the purpose
of the argument we will suppose that the body is what is called
a *solid of revolution*, i.e. it is a body such that all transverse
sections such as YY are circles, in other words it is the kind of
body that we could turn in a lathe, the axis of rotation being in
the section XX.

Now consider two equal portions,
each of weight w, opposite each other
and at the same distance from XX;
the moment of one about XX is wx
and of the other $-wx$ so that the
sum of the two moments about XX
is zero. As the whole body might
be divided up into similar neutralising
portions, the total moment of the
whole of the separate weights about
XX must be zero; in other words
XX must pass through the centre of
gravity. It will be noted that in the
case of the cylinder, G is what in
ordinary language we should call the
geometrical centre of the cylinder

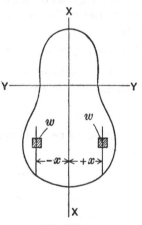

Fig. 117 a.

and in all similarly regular bodies the geometrical centre of
the body is the same as the centre of gravity. But if the

body be made of material of varying density or if there are blow-holes in it this will not be the case.

Numerical Example. *A uniform rod 24 inches long weighs 10 lbs. and carries at its ends balls of 4 and 6 inches diameter weighing respectively 5 and 8 lbs. Where is the centre of gravity ?*

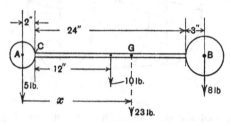

Fig. 118.

Referring to Fig. 118, the combined weight of 23 lbs. acts at the centre of gravity G, and since the separate bodies are symmetrical their weights act at their geometrical centres as shown.

We may take moments about any convenient point; take for example the centre A of the left-hand ball. Then we have

Moment of left-hand ball about $A = 5 \times 0$	$= \quad 0$
Moment of rod about $A = 10 \times 14$ (clockwise)	$= + 140$
Moment of right-hand ball about $A = 8 \times 29$ (clockwise) $= + 232$	
Total (in inch-lbs.)	$= + 372$

This must be equal to the moment of the total weight of 23 lbs. about A,

i.e. $$23x = 372,$$

$$x = \frac{372}{23} = \underline{16 \cdot 2 \text{ inches.}}$$

As a check the student should solve the problem by taking moments about the centre B of the other ball. It should be noted that there is no need to choose the centre of one of the separate bodies for taking moments, although that usually makes the

calculation easier.   For instance we might have chosen the point
C of junction of the rod with the left-hand ball   Then we have

Moment about C of left-hand ball = 5 × 2 (anti-

clockwise) = − 10

Moment about C of rod = 10 × 12 (clockwise)   = + 120

Moment about C of right-hand ball = 8 × 27 (clockwise) = + 216

Total (in inch-lbs.)   = + 326

$$\therefore \text{Dist. } GC = \frac{326}{23} = 14\cdot2 \text{ inches.}$$

This agrees with the previous result.

Centroid of an area.   There are a large number of engineering
problems in which we require to find the point in an area which
would be the centre of gravity of a thin uniform flat sheet of the
same contour as the area.   An area has no weight, so that it is
not strictly correct to speak of the centre of gravity of an area;
it is therefore called the *centroid*.   Many people, however, use the
term centre of gravity for both cases.

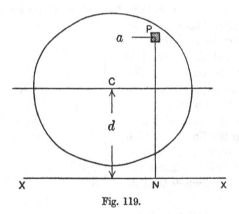

Fig. 119.

In the case of areas we may define the moment of an element
of area about a line as the product of the element by its perpendicular distance from the line.

Referring to Fig. 119, a is an element of area situated around
a point P, then $a \times PN$ is the moment of the element about XX.
If the whole area is divided up into elements and the moments of

the elements are added together, the result is called the moment of the whole area.

This is written

$$\text{Moment of area about } XX = \Sigma\,(a\,.\,PN).$$

From this point of view we may define the centroid as the point at which we can consider the whole area concentrated to give the same moment about any line.

If A is the area and it is considered as concentrated about the centroid C, then $A \times d$ is equal to the moment of area about XX;

$$\therefore\ d = \frac{\Sigma\,(a\,.\,PN)}{A} \ \ldots\ldots\ldots\ldots\ldots(5).$$

This is equivalent to the result we obtained in equations (3) and (4) for determining the position of the centre of gravity for solid bodies and we may take it that all the rules for finding the centre of gravity of a solid body can be applied to finding the centroid of an area.

Centroid of a triangle. Let ABD, Fig. 120, represent a triangle and let HJ be a very narrow strip drawn parallel to the base. Since this strip is so narrow it may be considered as a rectangle and its centroid is therefore at its centre K. The whole triangle may be considered divided up into strips, the centroid of each of which will be along the line AE which bisects the base at E and is called a *median line*. Therefore the centroid of the whole triangle must lie on AE.

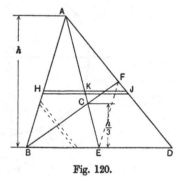

Fig. 120.

Similarly if we considered strips parallel to AD we should show that the centroid of the whole triangle must lie upon BF where F bisects AD.

The centroid of the triangle must therefore be at the intersection C of the median lines, and CE will be equal to $\dfrac{EA}{3}$; this is proved as follows. Join FE, then by a well-known geometrical property of the triangle, FE will be parallel to AB and will be equal to $\dfrac{AB}{2}$.

Therefore the $\triangle$s ABC, CFE are similar;

$$\therefore \frac{CE}{AC} = \frac{EF}{AB} = \frac{1}{2};$$

$$\therefore CE = \frac{AC}{2},$$

or $$CE = \frac{AE}{3}.$$

We get therefore the rule that "*the centroid of a triangle is along a median line and is at a distance from the base equal to one-third of the height.*"

Centre of gravity of a triangular pyramid. Let $ABDE$, Fig. 121, represent a triangular pyramid and let C_1 be the centroid

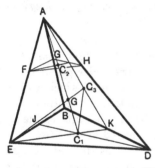
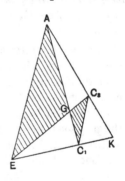

Fig. 121.

of the base. Join C_1A and consider a plane section FGH drawn parallel to the base BED. AC_1 cuts this section in C_2, and it can be proved by an application of the principle of similar triangles that C_2 is also the centroid of the $\triangle FGH$. Therefore the line AC_1 passes through the centroids of all the plane sections drawn parallel to the base so that the centre of gravity of the whole body must lie upon the line AC_1; similarly it must also lie upon EC_3, so that the centre of gravity of the body is at the intersection G of AC_1 and EC_3. Now consider the section AEK of the pyramid through the edge AE and the point K; this is shown on the right-hand side of the figure re-drawn for

14—2

greater clearness. It contains the point G. Now consider the $\triangle$s C_1C_3K and GAE;

$$\frac{C_3K}{AK} = \frac{1}{3} \text{ and } \frac{C_1K}{EK} = \frac{1}{3}$$

[from previous proof for the triangle].

Therefore the $\triangle$s are similar and C_1C_3 is parallel to AE and is equal to $\frac{AE}{3}$.

Next consider the $\triangle$s GC_1C_3 and GAE; their corresponding sides are parallel so that they also are similar;

$$\therefore \frac{GC_1}{AG} = \frac{C_1C_3}{AE} = \frac{1}{3};$$

$$\therefore GC_1 = \tfrac{1}{3}AG,$$

or $$GC_1 = \tfrac{1}{4}AC_1.$$

We see therefore that "*for a triangular pyramid the centre of gravity is on the line joining the apex to the centroid of the base and is at a height from the base equal to one-fourth of the height of the pyramid.*"

Extension to polygonal pyramid and cone. A polygonal pyramid (i.e. one whose base has more than three straight sides) may be divided up into a number of triangular pyramids of the same height the centre of gravity of each of which will be at one-fourth of the height of the base, so that the centre of gravity of the polygonal pyramid will also be at one-fourth of the height on a line joining the vertex to the centroid of the base.

A cone may similarly be considered as divided up into an infinite number of triangular pyramids with sides radiating from the centre of the base so that in the cone also the centre of gravity is at a distance from the base equal to one-fourth of the height.

Centroid of a trapezium. A trapezium is a four-sided figure with two sides parallel [some writers call it a "trapezoid"]. Referring to Fig. 122, if we considered narrow strips drawn parallel to the base, it is clear that the centroid of each must lie at the mid-point of each strip so that the centroid of the whole figure must lie somewhere upon the line FG joining the mid-points of the parallel sides.

Draw BJ parallel to the side AE. We then have the figure divided up into a parallelogram $ABJE$ and a triangle BJD, the centroids C_1 and C_2 of which will be at distances equal to $\dfrac{h}{2}$ and $\dfrac{h}{3}$ from ED. We now require to find the distance d of the centroid C of the whole figure from ED. We have seen already that it must lie on FG and it must also lie on the line C_1C_2 joining the centroids of the two parts.

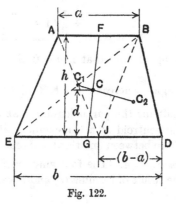

Fig. 122.

The area of the whole figure

$= $ Area of $ABJE$ + Area of BJD

$= ah + \dfrac{(b-a)h}{2} = h\left(a + \dfrac{b-a}{2}\right)$

$= \dfrac{h}{2}(2a + b - a)$

$= \dfrac{h}{2}(a + b)$ ....(1)

$= $ Half the height × Sum of the parallel sides.

Now take moments about the base ED.

Moment $= M = $ Moment of $ABJE$ + Moment of BJD

$= $ Area of $ABJE \times \dfrac{h}{2}$ + Area of $BJD \times \dfrac{h}{3}$

$= ah\cdot\dfrac{h}{2} + \dfrac{(b-a)h\cdot h}{2\cdot 3}$

$= h^2\left(\dfrac{a}{2} + \dfrac{b-a}{6}\right)$

$= \dfrac{h^2}{6}(3a + b - a)$

$= \dfrac{h^2}{6}(2a + b)$ ....(2).

But $\qquad M = $ Area of whole figure $\times d$

$$= \frac{h}{2}(a+b).d;$$

$$\therefore \quad \frac{h}{2}(a+b).d = \frac{h^2}{6}(2a+b),$$

i.e. $\qquad\qquad d = \frac{h}{3}\left(\frac{2a+b}{a+b}\right)$

$$= \frac{h}{3}\left(1 + \frac{a}{a+b}\right) \quad \ldots\ldots\ldots(3).$$

It is interesting to note that if $a = 0$, $d = \frac{h}{3}$, this being the case for the triangle; whereas if $a = b$, $d = \frac{h}{3}\left(1 + \frac{1}{2}\right) = \frac{h}{3}\cdot\frac{3}{2} = \frac{h}{2}$, which is the result for the parallelogram. As a check therefore we note that the centroid of a trapezium is at a distance from the base somewhere between one-third and one-half of the height.

Graphical construction. The following graphical construction is based upon the result of formula (3) and is very useful in many problems.

Set out BK, Fig. 123, $= b$ and $EH = a$ and join across as shown. Then the intersection of HK and FG gives the centroid C required.

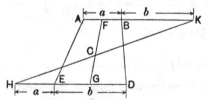

Fig. 123. Graphical construction for centroid of trapezium.

Graphical construction for centroid. The position of the centroid of any figure can be obtained by the following construction which is a special case of the link and vector polygon construction (p. 28).

Divide the area, Fig. 124, up into a number of small strips of equal breadth, parallel to the direction about which moments are taken, and draw the centre line of each of the said strips. Then if the strips are sufficiently small (we have only taken a few strips in the figure to avoid complication) the lengths of these centre lines represent the areas of the separate strips. Now, on a vector line, to some scale, set out 0 1, 1 2,...6 7 to represent the area of each strip, and take a pole P at convenient distance $= p$ from this vector line. Then anywhere across space 0 draw and produce a line ah parallel to $0P$; across space 1 draw ab parallel to $P1$; across space 2, bc parallel to $P2$, and so on until the point g is reached. Then draw the last link gh parallel to the last line $P7$ to meet ah in h.

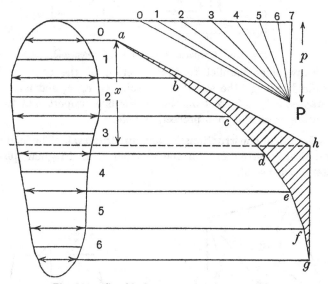

Fig. 124. Graphical construction for centroid.

Then the centroid lies on the dotted line through h drawn parallel to the given direction.

In most cases in practice we do not require the actual position of the centroid but only its distance from a line drawn in a given direction. In this case the above will suffice, the lines being drawn parallel to the given direction. If it does not, the construction should be repeated with the lines drawn at some convenient inclination—say at right angles—to the previous ones.

It is not really essential to divide the area into strips of equal breadth; any breadths may be taken, but in that case the *areas* of the strips must be set out on the vector line instead of the mid-ordinates.

When the figure can be divided up into a number of figures

the centroids of each of which can be found by inspection, we proceed as in Fig. 125.

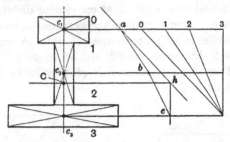

Fig. 125. Graphical construction for centroid.

Draw lines parallel to the direction of the centroid line required through the separate centroids c_1, c_2, c_3 and make the distances 0 1, 1 2, 2 3 on the vector line proportional to the separate areas and then proceed as before.

Graphical constructions for any quadrilateral. The following graphical constructions for the centroid of an irregular quadrilateral are useful.

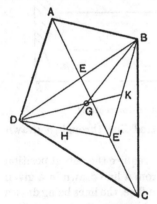

Fig. 126. Construction for centroid of quadrilateral.

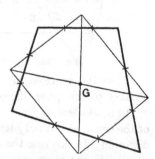

Fig. 127. Construction for centroid of quadrilateral.

First method. Let E be the point of intersection of the diagonals AC and BD, Fig. 126; from C set off $CE' = AE$ and join DE' and BE', then the centroid of the quadrilateral will be the same as that of the $\triangle BE'D$. Therefore bisect BE' and $E'D$ in K, H and join DK, BH; then their point G of intersection gives the required centroid.

Second method. Divide each of the sides into three equal parts and join across as indicated in Fig. 127. The resulting figure gives a parallelogram whose centre G is the centroid required.

Numerical Examples. (1) *Find the weight and centre of gravity of a cast iron body consisting of a cylinder 6 inches in diameter and 9 inches long, with a cone of the same diameter and 6 inches high standing on the top. Cast iron weighs ·26 lb. per cu. in.*

$$\text{The volume of the cylinder} = \frac{\pi d^2 h}{4} = \frac{\pi \times 6^2 \times 9}{4}$$

$$= 81\pi;$$

$\therefore$ Weight of cylinder $= 81\pi \times ·26 = 66$ lbs. nearly.

$$\text{Volume of the cone} = \frac{\pi d^2 h}{12} = \frac{\pi \times 6^2 \times 6}{12}$$

$$= 18\pi;$$

$\therefore$ Weight of cone $= 18\pi \times ·26 = 14·7$ lbs.;

$\therefore$ Total weight $= 66 + 14·7 = 80·7$ lbs.

The centre of gravity of the cone and cylinder are at G_1 and G_2 respectively, Fig. 128.

The centre of gravity G of the whole body will lie upon $G_1 G_2$ and regarding the separate weights as acting at right angles to the plane of the paper we can take moments about any convenient point, say G_2.

Then

$$80·7 \times GG_2 = 14·7 \times 6 + 66 \times 0;$$

$$\therefore \ GG_2 = \frac{14·7 \times 6}{80·7} = 1·1 \text{ inches.}$$

Therefore the centre of gravity is at $1·1 + 4·5 = 5·6$ inches from the base of the cylinder.

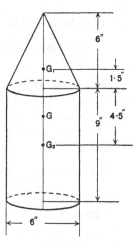

Fig. 128.

As an exercise the student should check this result by taking moments about G_1.

(2) *Find the position of the centroid of the cast iron beam section shown in Fig.* 129.

The centroid C is obviously upon the line of symmetry YY. To find its distance d from the base divide the section up into three rectangles as indicated, the area of each being regarded as acting at the centre.

Then we have total area

$$A = 2 \times 1\tfrac{1}{2} + 7 \times 1 + 6 \times 1\tfrac{1}{2}$$
$$= 19 \text{ sq. ins.}$$

Taking moments about the base we have

$$Ad = 3 \times 9\cdot25 + 7 \times 5 + 9 \times \cdot75$$
$$= 27\cdot75 + 35 + 6\cdot75$$
$$= 69\cdot5 \text{ in. units ;}$$
$$\therefore \ d = \frac{69\cdot5}{19} = 3\cdot66 \text{ inches.}$$

As an exercise the student should check this by the graphical construction shown in Fig. 125.

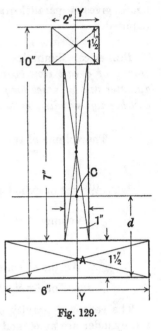

Fig. 129.

(3) *Find the position of the centroid of the angle section shown in Fig.* 130.

In this case we have a section which has no axis of symmetry and the centroid of which will lie outside the section. Divide up into two rectangles as shown.

$$\text{Total area} = A = 4\tfrac{1}{2} \times \tfrac{1}{2} + 3 \times \tfrac{1}{2}$$
$$= 2\cdot25 + 1\cdot5 = 3\cdot75.$$

Take moments about AB,

$$Ad_x = 1\cdot5 \times \cdot25 + 2\cdot25 \times 2\cdot75$$
$$= 6\cdot562 ;$$
$$\therefore \ d_x = \frac{6\cdot562}{3\cdot75} = 1\cdot75 \text{ ins.}$$

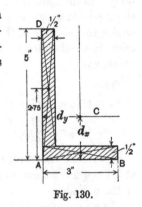

Fig. 130.

Take moments about AD,
$$Ad_y = 2 \cdot 25 \times \cdot 25 + 1 \cdot 5 \times 1 \cdot 5$$
$$= 2 \cdot 81;$$
$$\therefore \ d_y = \frac{2 \cdot 81}{3 \cdot 75} = \cdot 75 \text{ in.}$$

(4) *A circular disc 8 inches in diameter has cut out of it a circle of 3 inches diameter, leaving 2 inches on one side as indicated in Fig. 131. Find the centre of gravity of the resulting body.*

The area of the whole disc $= \dfrac{\pi \times 8^2}{4}$.

The area of the piece cut out $= \dfrac{\pi \times 3^2}{4}$.

$\therefore$ The relative weights are 8^2 and 3^2, i.e. 64 and 9.

$\therefore$ The relative weight of the remainder $= 64 - 9 = 55$.

The line XX is a line of symmetry so that the centre of gravity G lies upon it.

The centre of gravity of the whole disc is at G_1 and of the small circle at G_2.

Take moments about G_1.

Then
$$55 \cdot GG_1 = 9 G_1 G_2,$$
$$55 GG_1 = 9 \times \cdot 5,$$
$$GG_1 = \frac{9 \times \cdot 5}{55} = \underline{\cdot 082 \text{ in.}}$$

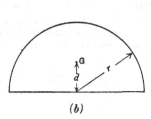

Fig. 131.

Centroid of various figures. The positions of the centroid of the following figures are useful in calculations, but their proof is beyond our present scope.

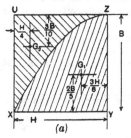

(a)

(b)

Fig. 132.

Parabola. Fig. 132 (*a*). The area of the interior segment $XYZ = \frac{2}{3}BH$ and of the exterior segment $XZU = \frac{BH}{3}$. The centroids G_1, G_2 are as indicated.

Semicircular arc. The centre of gravity of a rod bent to a semicircle will be at G, Fig. 132 (*b*), where $d = \frac{2r}{\pi}$.

Semicircular area. The centroid G is given by $d = \frac{4r}{3\pi}$.

Experiments upon centre of gravity and centroid. *Centre of gravity of a plate by suspension.* The centre of gravity of a plate or lamina can be found by hanging it up by one point *A*, Fig. 133, and drawing a line on the plate continuing the direction of the string, etc. The line then passes through the

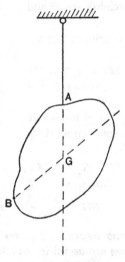

Fig. 133.

centre of gravity. The plate is then suspended from some other point *B*, preferably in about the relative position shown, and another similar line drawn. The intersection gives the centre of gravity *G*.

Centre of gravity of a walking stick. An interesting but very simple experiment can be performed with a walking stick as follows. Hold the stick horizontally with one finger near each end as indicated in Fig. 134 Then

Fig. 134.

move the fingers *A, B* towards each other fairly slowly without jerking and the fingers will meet at the centre of gravity.

After the student has finished reading this book, he should try to think out why this gives the centre of gravity.

Kinds of Equilibrium. Directly the line of pressure of a body falls outside the base, a moment acts which will make the body topple over; but if the line of pressure falls inside the base, the moment acting tends to maintain the body in equilibrium*.

It is common to speak of the equilibrium of a body as being one of three kinds, Fig. 135:

Stable equilibrium, in which the body tends to return to its original position of equilibrium when given a slight displacement.

Unstable equilibrium, in which the body tends to lose its equilibrium when given a slight displacement.

Neutral equilibrium, in which the body neither returns to its original position nor loses its equilibrium.

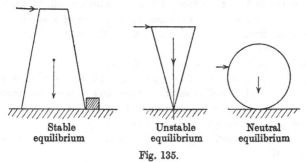

Stable   Unstable   Neutral
equilibrium   equilibrium   equilibrium

Fig. 135.

SUMMARY OF CHAPTER XIII.

The *centre of gravity* of a body is the point at which the resultant weight of the whole body may be considered to act.

$$X = \frac{\Sigma w_1 x_1}{\Sigma w_1} = \frac{\text{Sum of moments of separate weights}}{\text{Sum of separate weights}}.$$

The centre of gravity is the point about which a body will balance.

The centre of gravity of a body lies upon a section of symmetry.

The centroid of an area is the point at which the whole area may be considered concentrated to give the same moment about

* Cf. p. 22. The student should prove as an exercise that if the over-turning moment exceeds the stability moment, the line of pressure will fall outside the base.

any line. It is often called the centre of gravity, but strictly this is not correct because an area has no weight.

The centroid of a triangle is along a median line at a distance from the base equal to one-third of the height.

The centre of gravity of a pyramid or cone is on the line joining the apex to the centroid of the base and is at a height from the base equal to ¼ the height of the pyramid.

For a trapezium of height h, the distance of the centroid from the base b is $\dfrac{h}{3}\left(1+\dfrac{a}{a+b}\right)$, a being the side parallel to the base.

EXERCISES. XIII.

1. Find the position of the centroid of an isosceles triangle 4 inches base and 6 inches high.

2. Find the position of the centre of gravity of a cone 10 inches high and 8 inches diameter at the base.

3. A uniform rod of 5 lbs. is weighted with weights of 1 and 2 lbs. at the ends. Find the point about which it will balance.

4. Find the position of the centre of gravity of a square, length of side 2 ft., from which is cut out a circle of 1 in. diameter touching one of the sides at the centre.

Find the centre of gravity of the following:

5. A rod of length 2 feet weighing 2 lbs. to the end of which is fixed a spherical ball weighing 10 lbs. and 4 inches in diameter.

6. A T-shaped figure; the stem being 3 feet × 3 inches wide and the top 12 inches wide and 4 inches deep.

7. A balance weight having the form of a circular quadrant of radius R.

8. A trapezoidal wall 30 ft. high has a vertical back and sloping front face. The base is 10 ft. and top 7 ft. wide. What force must be applied horizontally at a point at 20 feet from the top to overturn it?   Take width of wall = 1 ft. and weight of masonry 130 lbs. per cu. ft.

9. A figure is made up of a square upon which stands an isosceles triangle. Find the relation between the height and base of the triangle in order that the centroid of the whole figure may be in the common base.

10. Find the position of the centroid of a channel section of base 10 inches, sides 3 inches and thickness of metal ¼ in.

11. Find the centre of gravity of the given figure. (See Fig. XIII a.)

12. Find the centre of gravity of an angle iron 4″ × 3″ × ½″.

13. Find the distance of the centre of gravity of the trapezium *ABCD* from *CD* (Fig. XIII *b*).

14. A rod 5 ft. long has a weight of 2 lbs. at one end and 3 lbs. at the other, also a weight of 5 lbs. at centre. Neglecting the weight of the rod, find the point about which it will balance.

15. A ship with equipment weighs 6000 tons. How far will its centre of gravity move if a gun weighing 30 tons is moved 20 feet across the deck?

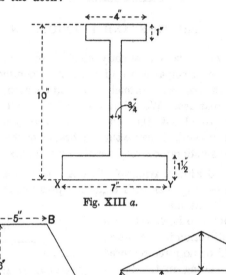

Fig. XIII *a*.

Fig. XIII *b*.   Fig. XIII *c*.

16. The bending moment of a beam of span l is made up of a triangle of height $\frac{pl^2}{16}$ at the centre and a parabola of height $\frac{pl^2}{32}$ extending from the right-hand end to the centre (Fig. XIII *c*). Taking the area of a parabola = $\frac{2}{3}$ base × height find the position of the centroid of the diagram from one end.

17. A solid cone 2 ft. high on a circular base has $\frac{1}{8}$ of its volume removed, being cut by a plane parallel to its base. Find the position of the centre of gravity of the remainder.

18. A circular disc 6 feet in diameter has a circular hole 6 inches in diameter cut out from it, the centre of the hole being 2 feet from one edge of the disc. How far will the centre of gravity be from the nearest edge?

CHAPTER XIV

FRICTION AND LUBRICATION

WE have explained already that a resistive force called *friction* is the principal cause of the loss of energy in machines and also that in some cases, such as in road traction, this frictional force is of great use. We will now consider the subject in greater detail and would ask the student to try to grasp fully each point as he proceeds, because this is a branch of the subject which is not always understood very clearly by students.

Static and kinetic friction. Let A and B, Fig. 136, be two bodies pressed together with a normal pressure P. Then since this force P has no component at right angles to itself (i.e. in a horizontal direction in the figure) there should be no force required to cause a sliding motion of A upon B. But actually there is a force tending to resist this sliding motion, and this resistive force is called the *force of friction*.

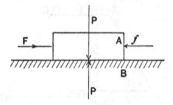

Fig. 136. Friction.

Now as the force F is slowly increased the resistive force f increases also, but we soon reach the condition when the sliding motion will commence because the force f is not capable of exceeding a certain value called the *limit of static friction*. The word static is used because the bodies are relatively stationary, and some writers have used the ingenious term *stiction* for it. As F gets still larger motion takes place and a frictional force f still comes into play, but it will not be quite equal to the limit of static friction and will depend to some extent upon the speed of sliding. The frictional force is then called *kinetic friction* or *friction of motion*.

Coefficient of Friction. The quantity $\dfrac{f}{P}$ is called the coefficient of friction and is generally given the letter μ.

$$\therefore\ \mu = \frac{f}{P} \quad\dotfill(1)$$

or
$$f = \mu P \quad\dotfill(2).$$

If therefore we are given the value of the coefficient of friction for the materials and conditions under consideration we can at once find the value of the friction force when the pressure between the surfaces is given.

With regard to static friction, the values of μ which are tabulated in the various books are those for the limiting friction. It should be remembered that the frictional force only becomes equal to μP at the moment when slipping is about to occur; until this condition is reached, the friction force f will be equal to the force F because if there is no relative motion between the two bodies the forces must be in equilibrium.

It has been found by experiment that for two given materials the coefficient of friction with dry surfaces is practically constant for various pressures; it is, however, a little smaller for a large pressure acting upon a small area than for a small pressure acting upon a large area.

Limiting reaction with friction; angle of friction. We have already seen that with smooth surfaces the reaction is always normal, i.e. at right angles to the surface*. With rough surfaces the reaction will be inclined to the normal in such a manner as to tend to oppose the motion. In the limiting condition when motion is just about to take place, the reaction reaches its limiting position and the angle at which it is inclined to the normal is called the *angle of friction.*

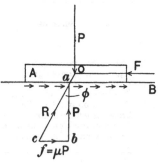

Fig. 137.   Angle of Friction.

We will explain this more fully with reference to a diagram, Fig. 137.

* p. 59.

The body A rests upon the body B and a force P acts in a direction at right angles to their surface of contact; a force F acts parallel to the surface and the body A is in equilibrium. Between the two surfaces there acts the friction force f indicated by the small arrows in the figure and a reaction pressure P which, in accordance with Newton's third law, is equal and opposite to the force P.

The reaction pressure P and the friction force f have a resultant reaction R which is inclined at an angle ϕ to the normal; this angle is called the *angle of friction*.

We may therefore define the angle of friction as follows:

The angle of friction is the angle with the normal which is made by the resultant reaction between two surfaces when slipping is about to take place.

Referring again to the figure we note that abc is a triangle of forces and that

$$\frac{bc}{ba} = \tan\phi = \frac{f}{P} = \frac{\mu P}{P} = \mu.$$

Therefore we obtain the rule that:

The tangent of the angle of friction is equal to the coefficient of friction.

Average values of μ. The following values of μ and ϕ may be taken as average values for dry surfaces.

Surfaces	Coefficient of friction μ	Angle of friction ϕ
Oak on oak (along grain)	·48	25° 38′
„ „ (across grain)	·34	18° 47′
Wrought iron on wrought iron	·14	8° 0′
Cast iron on cast iron	·15	8° 30′
Cast iron on oak (parallel to grain)	·49	26° 6′
Brass on wrought iron	·16	9° 6′
Common brick on common brick ..	·64	32° 30′
Masonry on moist clay...........	·33	18° 15′

Numerical Examples. (1) *A block weighing* 30 *lbs. rests upon a rough plate whose coefficient of friction is* ·2. *Find the least force acting horizontally which will move it.*

In this case $P = 30$ lbs. and $\mu = \cdot 2$,

$\therefore$ Limiting friction $= \mu P$

$= \cdot 2 \times 30$

$= 6$ lbs.

$\therefore$ Any force exceeding 6 lbs. will move the block.

(2) *In the above case, what is the least inclined force which will move the block and what will be its direction?*

We can solve this problem by considering the triangle of forces. Referring to Fig. 138, the three forces acting upon the block are

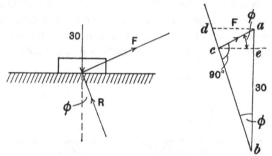

Fig. 138. Friction with inclined force.

the weight of 30 lbs., the resultant reaction R acting at an angle ϕ to the normal in a direction tending to oppose the motion and the tractive force F, whose direction we do not yet know.

We are given that $\mu = \tan \phi = \cdot 2$, and from trigonometrical tables we find that the value of ϕ is about $11° 20'$.

To draw the triangle of forces we set down ab to represent 30 lbs. to a convenient scale and then draw bd at an angle ϕ to it. Then if F were horizontal we should draw ad horizontally to fix the point d, and if F were in any other given direction we should draw from a a parallel to it. It will be clear that the force F is least when the distance from a to bd is the least possible, i.e. when ac is drawn at right angles to bd as shown.

If the triangle be drawn accurately to scale and ac be measured it will be found to be about 5·88 lbs.

By calculation we should say

$$\frac{F}{30} = \frac{ac}{ab} = \sin \phi,$$

$\therefore$ $F = 30 \sin 11° 20'$

$= 5\cdot 88$ lbs. approx.

To determine the direction of F draw ce horizontally, then $\angle\, ace = 90° - \angle\, cae = \phi$ [from $\triangle abc$].

We see therefore that the *best direction in which to pull a body along a rough surface is at an angle to the surface equal to the angle of friction.*

Rolling Friction. It is a fact of universal experience that it is easier to push an article provided with wheels than to push one without. This is often explained by saying that rolling friction is less than sliding friction, but such explanation does not get us much further. The exact nature of so-called rolling friction is not understood, but in a pure rolling motion there is no sliding motion at all and as frictional forces are solely brought into play by sliding there is no friction in pure rolling; on this

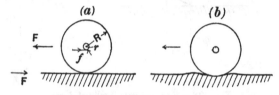

Fig. 139. Rolling Friction.

argument we should expect that with a very hard bed and roller the frictional resistance would be practically nothing. In most cases in practice, however, the roller or wheel sinks somewhat into the bed as indicated in Fig. 139 (b) and so the pure rolling action is stopped and some slipping occurs thus introducing friction. There is also a resistance due to the fact that the wheel is in a sense always going slightly uphill on account of the hump formed in the front of the depression. The harder the surface, the less will be the rolling resistance. Cyclists who have ridden upon very soft sandy roads and then come on to hard tarred roads will already have appreciated the truth of the above rule.

We should also expect the resistance to rolling to be less for wheels of large diameter than for small ones, because the smaller one will make a deeper depression than the larger one. Experiments show that this is true and that the following two rules are approximately true also:

(1) The rolling resistance is proportional to the load.

(2) The rolling resistance is inversely proportional to the diameter of the roller.

Action of wheels in assisting traction. In the case of a vehicle
such as a cart it might be argued that even if you have a wheel
you still have a sliding action at the axle which induces a friction
there instead of at the road. Against this we can point out that
the axle is iron running upon wood, brass or iron and that it is
easy to lubricate the axle; but this is not a complete answer.
The important point is that we have made the friction work at
a very small radius and the wheel gives a leverage over this
which makes the cart much easier to pull. Referring to Fig. 139 (*a*)
we see that the friction *f* acts at the axle circumference and the
tractive force *F* causes an equal resistance at the ground and this
force *F* has a large leverage over the friction force *f*.

Another way of looking at it is that in one revolution of the
wheel the work done against the friction is $f \times 2\pi r$ and the work
done by the traction is $F \times 2\pi R$. If the tractive force is just
sufficient to move the cart horizontally, these two amounts of
work will be the same;

$$\therefore f \times 2\pi r = F \times 2\pi R,$$
$$F = \frac{fr}{R}.$$

This is the same result as we should obtain by a consideration
of leverage.

It is important to remember that rolling resistance is small
only if the road is hard. It is easier to pull a flat bottomed box
along very soft sand than to pull the same box mounted on wheels.
It is an interesting fact that about 150 years ago wheeled carts
were practically unknown in the agricultural parts of Scotland
(see for instance Smiles' *Lives of the Engineers*); this was doubt-
less in part due to the fact that the roads were so bad that
practically all the advantage of wheels was lost.

Inclined plane and screw with friction. We have considered
already (p. 58) the case of the inclined plane (of which the
screw is a modification) in which frictional resistance was
neglected. We will now consider the case in which friction has
to be considered.

Case 1. *Force parallel to plane.*

(*a*) *Body moving upwards.* In this case the three forces
acting upon the body are the force *F*, Fig. 140, parallel to the
plane, the weight *W* vertically downward and the reaction *R*

which will be inclined at the angle of friction ϕ to the normal to the plane (shown in dotted lines).

We have already indicated (Fig. 33) that this normal is at an angle θ with the vertical, so that the reaction R makes an angle $(\theta + \phi)$ with the vertical. We are thus able to draw the triangle of forces abc, and the force F can either be found graphically or

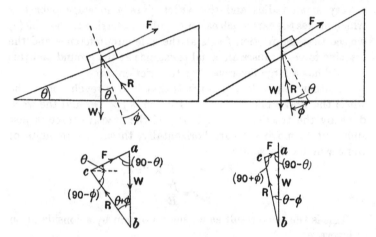

Figs. 140, 141. Inclined plane with Friction Force parallel to plane.

by calculation by means of a trigonometrical solution of the $\triangle abc$. Students who have gone sufficiently far with their trigonometry will understand the following solution:

$$\frac{F}{W} = \frac{ac}{ab} = \frac{\sin (\theta + \phi)}{\sin (90 - \phi)}$$

$$= \frac{\sin (\theta + \phi)}{\cos \phi};$$

$$\therefore F = \frac{W \sin (\theta + \phi)}{\cos \phi} \quad \dots\dots\dots\dots\dots(1),$$

$$\frac{R}{W} = \frac{bc}{ab} = \frac{\sin (90 - \theta)}{\sin (90 - \phi)}$$

$$= \frac{\cos \theta}{\cos \phi};$$

$$\therefore R = \frac{W \cos \theta}{\cos \phi} \quad \dots\dots\dots\dots\dots\dots(2).$$

(b) *Body moving downwards.* In this case the reaction still
acts at an angle ϕ to the normal, but on the opposite side of it,
so that it makes an angle $(\theta - \phi)$ with the vertical.

The triangle of forces *abc* is then as in Fig. 141 and by similar
reasoning we get

$$F = \frac{W \sin (\theta - \phi)}{\cos \phi} \quad \dots\dots\dots\dots\dots(3),$$

$$R = \frac{W \cos \theta}{\cos \phi} \quad \dots\dots\dots\dots\dots\dots(4).$$

It will be noted that R is the same in both cases.

Case 2. Force horizontal.

(a) *Body moving upwards.* In this case we have in Fig. 142
the triangle of forces *abc*, R being as before at an angle ϕ to the
normal.

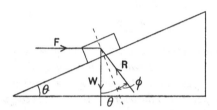 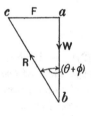

Fig. 142. Friction on inclined plane. Force horizontal.

We then have from the triangle *abc*

$$\frac{F}{W} = \frac{ac}{ab} = \tan (\theta + \phi);$$

$$\therefore \ F = W \tan (\theta + \phi) \quad \dots\dots\dots\dots(5),$$

$$\frac{R}{W} = \frac{bc}{ab} = \operatorname{cosec} (\theta + \phi);$$

$$\therefore \ R = W \operatorname{cosec} (\theta + \phi) \quad \dots\dots\dots\dots(6).$$

(b) *Body moving downwards.* In this case we shall have
as in Case 1 the reaction R still at an angle ϕ to the normal but
it will be on the opposite side, so that it will be inclined at an
angle $(\theta - \phi)$ to the vertical and we shall have by a similar
consideration of the triangle of forces

$$F = W \tan (\theta - \phi) \quad \dots\dots\dots\dots\dots(7),$$

$$R = W \operatorname{cosec} (\theta - \phi) \quad \dots\dots\dots\dots\dots(8).$$

Numerical Examples. (1) *A weight of 20 lbs. rests upon an inclined plane whose base is 4 feet long and whose height is 3 feet and is just prevented from moving downwards by a force of 8 lbs. acting horizontally. Find the coefficient of friction.*

In this case the body is about to move downwards and $F = 8$ lbs. and $W = 20$ lbs.

Therefore by equation (7)

$$8 = 20 \tan (\theta - \phi);$$
$$\therefore \tan (\theta - \phi) = \tfrac{8}{20} = \cdot 4.$$

A table of tangents shows us that $\tan 21 \cdot 8^c$ is approx. $\cdot 4$, so that we have $(\theta - \phi) = 21 \cdot 8°$.

Now $\tan \theta = \tfrac{3}{4} = \cdot 75$ and from tables we have $\theta = 36 \cdot 9°$ about;

$$\therefore \quad \phi = 36 \cdot 9 - 21 \cdot 8 = 14 \cdot 7° \text{ about,}$$
$$\therefore \quad \mu = \tan 15 \cdot 1°$$
$$= \cdot 27 \text{ about.}$$

(2)* *A block weighing 60 lbs. is on the point of motion down a rough inclined board when supported by a force of 24 lbs. acting parallel to the board and just begins to move up when acted upon by a force of 36 lbs. also parallel to the board. What is the coefficient of friction?* ·

In this case we have, when about to move down, by equation (3)

$$24 = \frac{60 \sin (\theta - \phi)}{\cos \phi},$$

i.e. $$\sin (\theta - \phi) = \frac{2 \cos \phi}{5} \quad \ldots \ldots \ldots \ldots (9).$$

When about to move up we have by equation (1)

$$36 = \frac{60 \sin (\theta + \phi)}{\cos \phi},$$

$$\therefore \sin (\theta + \phi) = \frac{3 \cos \phi}{5} \quad \ldots \ldots \ldots \ldots (10).$$

Now $$\sin (\theta - \phi) = \sin \theta \cos \phi - \cos \theta \sin \phi$$
and $$\sin (\theta + \phi) = \sin \theta \cos \phi + \cos \theta \sin \phi \quad \ldots (11).$$

---

* Students who do not possess fair knowledge of trigonometry will not be able to follow this example.

∴ putting these values in (9) and (10) and adding we get

$$2 \sin \theta \cos \phi = \cos \phi \, ;$$
$$\therefore \ \sin \theta = \tfrac{1}{2},$$
or
$$\theta = 30° \, ;$$
$$\therefore \ \sin(\theta + \phi) = \sin 30 \cos \phi + \cos 30 \sin \phi$$
$$= \frac{\cos \phi}{2} + \frac{\sin \phi \sqrt{3}}{2}.$$

∴ by (10)
$$\frac{3 \cos \phi}{5} = \frac{\cos \phi}{2} + \frac{\sin \phi \sqrt{3}}{2},$$

i.e.
$$\frac{\sin \phi \sqrt{3}}{2} = \cos \phi \left(\frac{3}{5} - \frac{1}{2} \right) = \frac{\cos \phi}{10},$$

$$\therefore \ \frac{\sin \phi}{\cos \phi} = \frac{1}{5 \sqrt{3}} \, ;$$

$$\therefore \ \tan \phi = \mu = \frac{1}{5 \sqrt{3}} = \frac{\sqrt{3}}{15} = \cdot 115,$$

i.e.
$$\mu = \cdot 115.$$

Angle of Repose. The largest angle of an inclined plane upon which a body can rest without sliding down is called the angle of repose, and we can show in the following manner that the angle of repose is equal to the angle of friction.

When a body is just about to slide down, the force acting either parallel to the plane or horizontally is zero, so that by equation (7) we have

$$0 = W \tan(\theta - \phi).$$

Since W is not zero, $\tan(\theta - \phi)$ must be zero, i.e. $(\theta - \phi) = 0$ or $\theta = \phi$.

The angle of repose is of importance in considering the stability of walls supporting banks of earth.

The efficiency of a screw. We have shown on p. 63 that a screw is really a special case of an inclined plane with a horizontal force, and when friction was neglected we had the relation

$$F = W \tan \theta \quad \dots\dots\dots\dots\dots (12).$$

When friction is considered we get the following treatment:

(a) *Screwing in.* When screwing in, the load is moving up the plane, so that equation (5) is the one to use.

We have therefore

$$F = W \tan(\theta + \phi) \quad \dots\dots\dots\dots (5).$$

Now the efficiency (7) of a machine may be determined by the relation

$$\eta = \frac{\text{Ideal effort}}{\text{Actual effort}}$$

$$= \frac{W \tan \theta}{W \tan (\theta + \phi)}$$

$$= \frac{\tan \theta}{\tan (\theta + \phi)^*} \quad \ldots\ldots\ldots\ldots (13).$$

It can be shown that this efficiency is the maximum possible when $\theta = 45° - \dfrac{\phi}{2}$, but the proof of this is beyond our present standard.

(b) *Screwing out.* When screwing out, the load is moving down the plane, so that equation (7) is relevant;

$$\therefore \quad F = W \tan (\theta - \phi).$$

In this case F is the horizontal force that the weight W will move downwards, the effort and resistance being reversed.

$$\therefore \quad \text{Actual effort} = \frac{F}{\tan (\theta - \phi)},$$

$$\text{Ideal effort} = \frac{F}{\tan \phi};$$

$$\therefore \quad \eta = \frac{\text{Ideal effort}}{\text{Actual effort}}$$

$$= \frac{\tan (\theta - \phi)}{\tan \theta} \quad \ldots\ldots\ldots\ldots (14).$$

This is found to be a maximum when

$$\theta = 45° + \frac{\phi}{2}.$$

It will be noticed that if equation (14) = 0, it means that the screw will not run backwards unless it is helped round. The screw is then called self-locking and this in many machines is a useful feature but it means that the efficiency of the machine is sacrificed to it.

This occurs when $\theta = \phi$ or *if the angle of a screw is less than the angle of friction the screw will not run backwards*, i.e. *the nut will not drive the screw.*

$$* \tan (\theta + \phi) = \frac{\tan \theta + \tan \phi}{1 - \tan \theta \tan \phi}.$$

Ladder resting against a wall. If a ladder AB, Fig. 143, rest against a wall at B and against the ground at A, frictional forces are induced at A and B preventing the ladder from sliding down the wall.   If slipping is about to take place, the reactions R_A and R_B will be inclined as shown at angles ϕ_1 and ϕ_2 to the normal, ϕ_1 being the angle of friction for the ladder on the ground and ϕ_2 for the ladder along the wall.

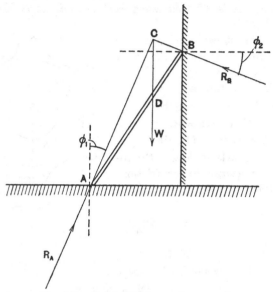

Fig. 143. Ladder resting against a wall.

Now we have already proved (p. 26) that if three forces are in equilibrium they must, if not parallel, pass through a point. As therefore R_A and R_B meet at C, the resultant weight W, of the ladder and of a man standing upon it, must also act through C.

It must always be remembered that in these friction problems it is only when slipping is just about to take place that the reactions are inclined at the angle of friction.   In other cases, as for instance when the resultant weight of the ladder comes below the point D, the reactions will be less inclined and their actual values cannot always be determined.   In the ladder problem, all that we know is that the reactions must intersect on the vertical line through the resultant weight and that neither reaction can be inclined to the normal at an angle greater than the angle of friction but there are a very large number of reactions possible which will satisfy these conditions. Problems of this kind in which the exact result cannot be found are called "statically indeterminate."

Numerical Examples. (1) *A wheel rotates upon an axle 3 inches in diameter and makes* 90 *revolutions per minute. If the load on the wheel is* ¼ *ton and the coefficient of friction for the lubricated axle is* ·02, *how much work per minute is absorbed in friction ?*

Fig. 144 shows the axle, the weight being regarded as acting at the bottom.

In this case the load

$$= W = \tfrac{2240}{4} = 560 \text{ lbs.} ;$$

∴ Friction force $= \mu W$

$$= \cdot 02 \times 560 = 11 \cdot 2 \text{ lbs.}$$

Now the axle is constantly rotating in opposition to this friction.

∴ Distance moved per minute

$$= \pi D N$$

$$= \frac{\pi \times 3 \times 90}{12} \text{ ft.}$$

$$= 70 \cdot 7 \text{ ft.}$$

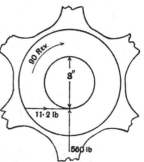

Fig. 144.

∴ Work done against friction per minute

$$= \text{Force} \times \text{Distance moved}$$

$$= 11 \cdot 2 \times 70 \cdot 7$$

$$= \underline{792 \text{ ft.-lbs.}}$$

We have taken the weight as acting at the bottom in this case because this is approximately true; strictly the weight will act a little to the right of the bottom, sufficiently far away for the resultant of the normal reaction and the friction force to be exactly equal and opposite to the weight.

(2) *Find the efficiency of a screw* 2½ *inches in diameter in which there are four threads to the inch and the coefficient of friction is* ·04.

In this case $p = \tfrac{1}{4}$ inch and the circumference $= 2 \cdot 5\pi$;

$$\therefore \ \tan \theta = \tfrac{1}{4} \div 2 \cdot 5\pi = \frac{1}{10\pi} = \cdot 032 \text{ approx.}$$

Now by equation (13)

$$\eta = \frac{\tan\theta}{\tan(\theta + \phi)} = \frac{\tan\theta\,(1 - \tan\theta\,.\,\tan\phi)}{\tan\theta + \tan\phi}$$

$$= \frac{\cdot032\,(1 - \cdot0013)}{\cdot032 + \cdot04} = \underline{\cdot44}\,;$$

∴ Efficiency of screw = ·44 or 44 %.

(3) *A cylinder weighing 6 lbs. is 2 inches in diameter and 8 inches long. It is placed on a board which is slowly tilted up. If the coefficient of friction between the board and the cylinder is ·2, will the cylinder start sliding before it topples over?*

We have already seen (p. 221) that if the line of action of the weight of a body falls outside the base it will topple over unless held down by some external means; we have also learnt that the body will start sliding when the slope is equal to the angle of friction. In Fig. 145 we have shown the board at this slope ($\mu = \tan\phi = \frac{2}{10} = \cdot2$); we therefore require to find whether

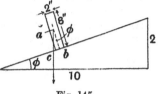

Fig. 145.

in this position the line of action of the weight of the cylinder falls outside the base. This can be done by drawing carefully to scale and then drawing a vertical through the centre of the cylinder. It will be found to come just inside so that the cylinder will slide before it topples over.

We can obtain this result by calculation as follows: the angle between *ab* and *ac* will also be ϕ;

$$\therefore \quad \tan\phi = \frac{bc}{ab}, \text{ but } ab = 4 \text{ inches,}$$

$$\therefore \quad \cdot2 = \frac{bc}{4},$$

or $bc = 4 \times \cdot2 = \cdot8$ inch.

As the cylinder is 2 inches in diameter the distance from *b* to the edge of the base will be 1 inch so that *c* falls inside the base.

Lubrication. The purpose of lubrication is to reduce friction and so minimise the energy which is absorbed by the friction. This is effected by imprisoning a film of oil between the two surfaces so that the friction between the surfaces is replaced by a friction between the fluid and the surfaces and this is less than the friction between the dry surfaces.

In the choice of a lubricant it should be remembered that the condition under which it is to be used should be considered. It should have sufficient viscosity to prevent its being squeezed right out of the bearing and if the part lubricated is likely to be hot in working the lubricant should be such that its lubricating properties are not destroyed at the higher temperature. In designing lubricating devices care should be taken that the lubricant is not introduced at the point where the pressure is greatest; otherwise little will find its way to the bearing. The friction of lubricated bearings is really a subject requiring separate attention and it is rather beyond our present stage.

Experiments upon Friction. The following experiments can be made with very simple apparatus.

(1) *Determination of the coefficient of friction by tilting.* Hinge a board *C* at one end to a board *A* (Fig. 146), and at each side of the opposite end of the

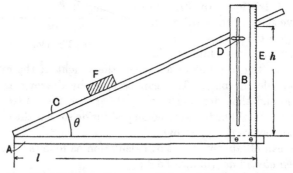

Fig. 146. Determination of coefficient of friction by tilting.

latter set up two slotted uprights *B*. Between the uprights fix a bolt *D* provided with a fly-nut by means of which it can be fixed in any position, upon which the board *C* can rest. On the edge of the upright *B* fix a scale *E*. The block *F* whose angle of friction with the board *C* is required is placed upon the board *C* which is slowly tilted upwards until the block begins to slide.

The height *h* at which sliding commences is noted and then we have, as shown on p. 226,

$$\mu = \tan \theta = \frac{h}{l}.$$

By choosing *l* a convenient round number of inches, the scale *E* can easily be graduated to read off values of μ direct. To make an instructive experiment blocks *F* of different weights and areas of contact for the same material may be taken so that the student can discover for himself what effect the pressure has upon the coefficient.

(2) *Determination of the coefficient of friction by weights.* Fix a smoothly running pulley B (Fig. 147) in the end of a board A and connect a thin string to a block D, the string passing over the pulley and having a scale pan attached to its end. A weight is then placed on the block, the combined weight including that of the block being W; small weights are then placed carefully in the scale pan until the block begins to slide. Then if f is the sum of the added weights and the weight of the scale pan, the limiting coefficient of static friction will be given by

$$\mu = \frac{f}{W}.$$

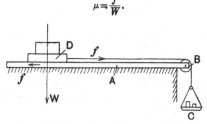

Fig. 147. Determination of coefficient of friction by weights.

The experiment may be extended by varying the weight W and by placing surfaces of various kinds upon the board A and also by turning the block D round to vary the direction of the grain.

Experiments may also be made to find the coefficient of kinetic friction by loading the scale pan until when the block is given a start it will continue to move uniformly.

(3) *Experiments upon rolling friction.* The same apparatus may be employed for experiments upon rolling friction by replacing the block D by a small model wheeled truck. The friction of the axles will of course come into play but the effect of various surfaces upon the friction for the same truck may be investigated by laying various surfaces upon the board.

SUMMARY OF CHAPTER XIV.

Friction is the force between two surfaces which tends to prevent them from moving relatively to each other.

When motion is already taking place the force is called *kinetic friction*, but when the bodies are stationary the friction force when sliding is just about to take place is called the *limit of static friction.*

$$f = \mu P,$$
$$\mu = \tan \phi.$$

Friction is reduced by replacing a sliding motion by a rolling motion.

To deal with friction on an inclined plane treat the reaction as acting at an angle ϕ to the normal to the surface on the side which will tend to oppose motion.

The angle of repose is equal to the angle of friction.

$$\text{Efficiency of screw} = \frac{\tan \theta}{\tan (\theta + \phi)}.$$

EXERCISES. XIV.

1. A weight of 5 cwt. resting on a horizontal plane requires a horizontal force of 100 lbs. to move it against friction. What in that case is the value of the coefficient of friction?

2. A body weighing 40 lbs. rests on a rough horizontal plane whose coefficient of friction = 0·25. Find the least horizontal force which will move the body.

3. A locomotive weighs 65 tons of which 0·48 of the whole rests on the driving-wheels. What must the coefficient of friction be between the driving-wheels and the rails so that the engine may draw a train of total weight 200 tons at 50 miles an hour up an incline 1 in 300? Resistance = 45 lbs. per ton.

4. A horse drags a load of 35 cwt. up an incline of 1 in 20. The resistance on the level is 100 lbs. per ton. Find the pull on the traces when they are (a) horizontal; (b) parallel with the incline; (c) in the position of the least pull.

5. If the angle of friction is 10°, find the magnitude and direction of the least force which will push a load of 20 tons up a plane inclined at 20° to the horizon.

6. A bicycle and rider weighing together 180 lbs. are going along the level at 10 miles an hour. If the brake be applied at the top of the front wheel (30″ diam.) and is the only resistance acting, how far will the bicycle travel before stopping if the pressure of the brake is 20 lbs. and $\mu = 0·5$?

7. Prove that a train going 60 miles an hour can be brought to rest in 313 yds. (about) by the brakes supposing them to press on wheels with ⅔ weight of the train and $\mu = 0·18$ in addition to a passive resistance of 20 lbs.-wt. per ton on the level.

8. A wheel 12 ft. in diameter, rotating at the rate of 1 rev. in 2 seconds, is acted upon by a brake which applies normal pressures of 1 cwt. each at opposite ends of a diameter. If $\mu = 0·6$, find the H.P. absorbed.

9. If the coefficient of friction be $\frac{3}{4}$, find the least depth from back to front of a drawer 2 ft. wide, which can be drawn out by a direct pull on a handle 6 ins. to the right or left of the middle of the front.

10. A ship weighing 2000 tons is launched. Find what slope of the ways is necessary for uniform motion when once started. Also what should be the area of the bearing surface so that the pressure shall not exceed $2\frac{1}{2}$ tons per sq. ft. and so force out the tallow. $\mu = 0\cdot14$.

11. In a screw-jack the pitch of the square-threaded screw is $0\cdot5$ in. and the mean diameter is 2 ins. The force exerted on the bar used in turning the screw is applied at a radius of 21 ins. Find this force if a load of 3 tons is being raised. Taking $\mu = 0\cdot2$, what is the efficiency of this machine?

12. A uniform ladder 70 ft. long is equally inclined to a vertical wall and horizontal ground, both of which are rough. The weight of a man and his burden ascending the ladder is 2 cwt. and the weight of the ladder is 4 cwt. How far up may he ascend before the ladder begins to slip if $\mu = \frac{1}{3}$ for the ground and $\frac{1}{2}$ for the wall?

CHAPTER XV

MOTION IN A CURVED PATH

The Hodograph. We have considered up to the present only the cases in which the motion of a body takes place in a straight line. If a body moves in a curved path, its motion may in many cases be considered most conveniently by means of the *Hodograph* which is defined as follows:

Let P_0, P_1 ... P_4, etc., Fig. 148, represent successive points upon the curved path of a body and let $P_0 0$, $P_1 1$, etc. be the tangents to the curve at the various points.

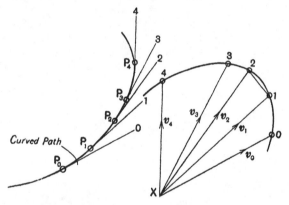

Fig. 148. The Hodograph.

Taking a pole X draw a vector $X0$ parallel to the tangent to represent the velocity v_0 of the body at the point P_0 in magnitude and direction to some convenient scale; then draw $X1$ parallel to $P_1 1$ to represent the velocity v_1 at P_1 to the same scale and so on. Then the curve obtained by joining the points 0, 1, 2 ... 4, etc. is called the *velocity hodograph* for the motion.

Now consider the question of acceleration. Acceleration is

defined as the rate of change of velocity, and the change in velocity may consist of a change of direction as well as one of magnitude. In the case under consideration for instance the velocity between the points P_1, P_2 changes from $X1$ to $X2$, the change in velocity being represented by the vector difference 1, 2. If the distance $P_1 P_2$ is very short and the time taken in traversing it is δt, we have

$$\text{Acceleration} = \frac{\text{Change in velocity}}{\text{Time taken}},$$

i.e.
$$a = \frac{1, 2}{\delta t} \quad \dots\dots\dots\dots\dots\dots\dots(1).$$

This means that the acceleration of the body between the points P_1, P_2 is equal to the velocity with which the corresponding point in the hodograph moves across the corresponding period.

This gives us the rule that "*the velocity in the hodograph is equal to the acceleration in the curved path.*" The acceleration at any point will also be in the direction of the tangent to the hodograph at the point.

If therefore we consider the velocity hodograph as a curved path and repeat the construction, the new curve will give accelerations and may be called the *acceleration hodograph*.

Uniform motion in a circle; angular velocity. Suppose that a point moves with a velocity v in a circle of radius r, and that in a time t the point moves through an arc AB, Fig. 149, subtending an angle θ at the centre of the circle.

Then the angle turned through in a unit time is called the *angular velocity* and is given the letter ω.

Then since

arc = angle (in radians) × radius

we have $\quad AB = r\theta$,

and if t is the time taken from A to B,

$$AB = vt;$$

$$\therefore \quad r\theta = vt$$

or $\qquad v = \dfrac{r\theta}{t},$

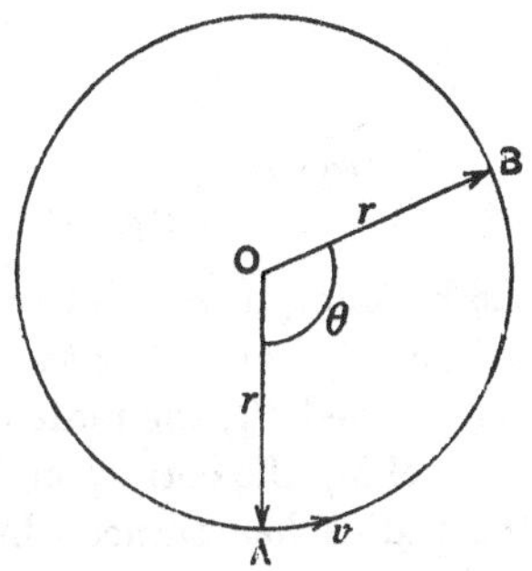

but $\dfrac{\theta}{t}$ = angular velocity = ω;

$$\therefore \quad v = \omega r \quad \dots\dots\dots\dots\dots\dots(2).$$

Fig. 149.

16—2

In practice angular velocity is not expressed in radians per minute or per second, but in revolutions per minute or per second.

Now in one revolution the point moves through a distance $2\pi r$ so that if a point rotates uniformly at N revolutions per second, the velocity at a radius r is given by

$$v = 2\pi r N \quad \dots\dots\dots\dots\dots\dots(3).$$

Numerical Example. *If a shaft 4 inches in diameter rotates at a uniform rate of 80 revolutions per minute, what is the peripheral velocity of the shaft in feet per second ?*

In this case $r = 2$ ins., $N = \frac{80}{60}$ per second.

∴ Peripheral velocity v in inches per second

$$= 2\pi r N$$
$$= 2 \times 3{\cdot}1416 \times 2 \times \tfrac{80}{60}$$
$$= 16{\cdot}76.$$

∴ Peripheral velocity $= \dfrac{16{\cdot}76}{12}$

$$= 1{\cdot}40 \text{ feet per second.}$$

Centripetal and centrifugal force. If a body moves with uniform velocity v feet per second in a circle of radius r feet (Fig 150), the velocity hodograph will be a circle of radius v, the radius $X0$ of

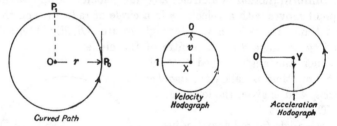

Fig. 150. Centripetal Acceleration.

the hodograph being at right angles to the corresponding radius OP_0 of the curved path. When the point in the curved path has reached P_1, the radius has turned through a right angle and, in reaching the corresponding point 1 on the velocity hodograph, the radius has turned through the same angle. The velocity hodograph therefore turns through a complete circle in the same time as the body moving in the curved path completes its circle.

The acceleration hodograph will also be a circle because it is obtained from the velocity hodograph by the same construction as that employed for drawing the former. The radius $Y0$ in the acceleration hodograph is parallel to the tangent at O and is thus at right angles to $X0$ and opposite to the radius OP_0 of the curved path; a revolution of the acceleration hodograph will also be completed when one revolution in the curved path is completed.

We get therefore the result that with uniform motion in a circle there is a constant acceleration towards the centre. This acceleration is usually called the *centripetal acceleration*; its magnitude can be found as follows.

Let t seconds be the time taken to complete the circle, then we have

$$v = \frac{2\pi r}{t} \quad \dots \dots \dots \dots \dots \dots (4);$$

also the acceleration a is the velocity on the velocity hodograph,

$$\therefore \; a = \frac{2\pi v}{t} \dots \dots \dots \dots \dots \dots (5).$$

Dividing we get

$$\frac{v}{a} = \frac{r}{v}$$

or

$$a = \frac{v^2}{r} \quad \dots \dots \dots \dots \dots \dots (6).$$

If the weight of the body is W, we have by the rule

$$\text{Force} = \frac{\text{Weight} \times \text{acceleration}}{g}$$

a constant "centripetal force" acting towards the centre of the circle to maintain the motion.

The force equal and opposite to this, which is the apparent force acting outwards upon the body, is called the "centrifugal force," the two terms being often confused.

The centrifugal force is really the force acting outwards at the same radius as the rotating body which will equilibrate or balance the system of forces acting on the body, as the examination of the equilibrium of such bodies is correctly dealt with by considering the forces acting on the body together with the reversed radial accelerating force as forming a system in equilibrium.

Since the weight of a body acts at the centre of gravity and the centrifugal force acting on each portion of the body is proportional to the weight of that body, it follows that the resultant centrifugal force also acts through the centre of gravity.

We get therefore, from equation (6),

Centripetal or centrifugal force

$$= F_c = \frac{Wa}{g}$$
$$= \frac{Wv^2}{gr} \quad \ldots\ldots\ldots\ldots\ldots(7).$$

If the velocity is given in terms of revolutions per minute (N)

we have, since $v = \frac{2\pi r N}{60}$,

$$F_c = \frac{4\pi^2 N^2 r^2 W}{3600 gr}$$
$$= \frac{4\pi^2 N^2 r W}{3600 g} \quad \ldots\ldots\ldots\ldots(8)$$
$$= \cdot00034 N^2 r W \quad \ldots\ldots\ldots\ldots(9).$$

Numerical Examples. (1) *What force acting horizontally tends to overturn a train weighing* 100 *tons when running round a curve of* 500 *feet radius at* 60 *miles per hour?*

In this case $W = 100$ tons,

$v = 60$ miles an hour

$= 88$ ft. per second,

$r = 500$ ft.

∴ Centrifugal force which tends to overturn the train

$$= F_c = \frac{Wv^2}{gr}$$
$$= \frac{100 \times 88 \times 88}{32 \cdot 2 \times 500}$$
$$= \underline{48 \text{ tons nearly.}}$$

(2) *At how many revolutions per minute must a stone weighing* ¼ *lb. whirl horizontally at the end of a string* 5 *feet long to cause a tension of* 2 *lbs. in the string?*

In this case $W = \frac{1}{4}$,

$F_c = 2$,

$r = 5$.

∴ using equation (9)

$$2 = \cdot00034 N^2 \cdot 5 \cdot \tfrac{1}{4},$$
$$N^2 = \frac{8}{5 \times \cdot00034},$$
$$N = \underline{68\cdot6 \text{ revolutions per minute.}}$$

Applications of centrifugal and centripetal force. There are a large number of problems in engineering practice in which centrifugal and centripetal force are of importance.

Railway Curves and Motor Tracks.

When a railway train or motor car goes round a curve the radial acceleration induces forces which tend to overturn it, and this has been the assigned cause of accidents even in recent years —for instance the railway accident at Salisbury a few years ago. Those students who have played with model steam-engines will have found that when the speed gets high the engine will often fall over at a bend.

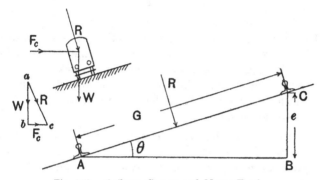

Fig. 151. Railway Curves and Motor Tracks.

To minimise these dangers it is now the practice to tilt or "super-elevate" the rails and to bank the motor track at a bend, the arrangement in the latter case being that the surface is perpendicular to the resultant of the weight of the body and the centrifugal force. It is commonly stated that in railway tracks also the surface should be perpendicular to this resultant, but such is not the case. In the railway track the problem is to give an elevation which will prevent the inner wheel from lifting off the rail; this means that the resultant of the weight and centrifugal force must act inside the tread of the outer rail. Referring to Fig. 151, the resultant force R is obtained by considering the triangle of forces abc; if the track is so banked up that this resultant acts at right angles to it, there will be no tendency for the body to overturn.

Our problem therefore becomes that of determining the angle θ so that AC is perpendicular to R.

Now each side of the $\triangle$ ABC is perpendicular to a side of the $\triangle$ abc; therefore these triangles are similar and

$$\frac{BC}{AB} = \frac{bc}{ab},$$

but

$$bc = F_c = \frac{Wv^2}{gr},$$

where v is the velocity of the train or car and r is the radius of the bend; also

$$ab = W,$$

$$\therefore \frac{BC}{AB} = \frac{Wv^2}{Wgr} = \frac{v^2}{gr},$$

i.e.

$$\tan \theta = \frac{v^2}{gr} \dots\dots\dots\dots\dots(10).$$

This result gives us the angle of tilt that should be provided to bring the resultant force at right angles to the surface and can be correctly applied to the case of the motor track.

This treatment would however, as the following numerical example shows, give a much higher super-elevation for railways than is ever adopted.

Numerical Example. *If the gauge of a railway is 4' 8½", find the super-elevation required for a curve of 400 feet radius at a speed of 60 miles an hour if the resultant force is to be perpendicular to the rails.*

In this case

$$v = 88 \text{ feet per second,}$$
$$G = 56\cdot5 \text{ inches,}$$
$$r = 400 \text{ feet,}$$
$$g = 32\cdot2 \text{ feet per second per second,}$$
$$\tan \theta = \frac{88 \times 88}{32\cdot2 \times 400}$$
$$= \cdot6012;$$
$$\therefore \theta = 31\cdot0° \text{ approx.;}$$
$$\therefore e = G \sin \theta$$
$$= 56\cdot5 \times \sin 31\cdot0°$$
$$= \underline{29\cdot1 \text{ inches nearly.}}$$

Centrifugal governors or conical pendulums. The centrifugal governor is a device for regulating the speed of engines and motors and in its simplest form was employed by James Watt. A simple form, shown in Fig. 152, has two balls carried by arms pivoted to a collar A upon a shaft O driven from the main shaft of the machine. The arms carry links pivoted to a sleeve B which is movable up and down the shaft O, the motion being transmitted by a bell-crank lever C to a rod D connected to the throttle valve of the steam-engine.

Should the speed of the engine increase, the radial force will increase and the balls will fly outwards and the sleeve B will rise and thus cut off the supply of steam until the engine has regained its normal speed.

We can find the relation between the height h and the radius r of the balls for any given speed.

The forces acting upon each ball are its weight W—called the "Controlling Force"—and the centrifugal force F_c.

Since the arms are freely pivoted to the collar A, the arms will move until there is no tendency to move about the pivot. But we have seen that the tendency of a number of forces to rotate a body about

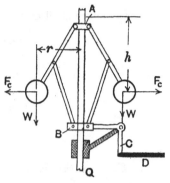

Fig. 152. Watt's Centrifugal Governor.

any point is measured by the sum of the moments of the forces about the point, so that in the present case this moment must be zero, and neglecting the weight of the arm we therefore have

$$F_c h - Wr = 0, \text{ i.e. } F_c h = Wr,$$

or
$$\frac{h}{r} = \frac{W}{F_c},$$

$$\frac{h}{r} = \frac{Wgr}{v^2} \text{ [by equation 7]},$$

$$\therefore h = \frac{gr^2}{v^2} \dots \dots \dots \dots \dots (11).$$

If we wish to use a formula in terms of the number of revolutions N per minute we use equation (8),

$$\therefore \frac{h}{r} = \frac{3600g}{4\pi^2 N^2 r},$$

i.e.
$$N^2 = \frac{3600gr}{4\pi^2 rh},$$

i.e.
$$N = \frac{60}{2\pi}\sqrt{\frac{g}{h}} \dots \dots \dots \dots (12).$$

Numerical Example. *Find the speed at which a simple centrifugal governor will run when the height is 9 inches and find the amount by which the balls will rise when the number of revolutions per minute increases by 5.*

In this case h = 9 inches = ·75 foot;

$$\therefore N = \frac{60}{2\pi} \sqrt{\frac{32 \cdot 2}{\cdot 75}}$$

= 63 revolutions per minute nearly.

If N = 66 we shall have

$$66 = \frac{60}{2\pi} \sqrt{\frac{32 \cdot 2}{h}},$$

$$\therefore 66^2 = \frac{60^2 \times 32 \cdot 2}{4\pi^2 h};$$

$$\therefore h = \frac{60 \times 60 \times 32 \cdot 2}{66 \times 66 \times 4\pi^2}$$

= ·674 foot nearly

= 8·09 inches.

$\therefore$ The balls rise by 9 − 8·09

= ·91 inch.

Balancing rotating parts. If a wheel or other revolving body has its centre of gravity out of the centre of rotation, then the whole body may be considered as a weight concentrated at its centre of gravity and thus rotating in a circle whose radius is equal to the distance from the centre of gravity to the axis of rotation.

The resulting radial force may at high speeds cause severe vibrations and will interfere with smooth running besides causing heavy stresses upon the shaft and bearings.

Rotating bodies which give rise to these centrifugal forces are said to suffer from want of balance and the problem of removing these forces and similar forces caused by rotating parts is called "balancing." In many cases this problem is an exceedingly difficult one, and in some cases of engines in electric power stations which caused severe vibrations in adjoining buildings great expense and inconvenience have resulted, due to the inability of even the leading authorities to quite remove the lack of balance.

We cannot at the present stage go fully into the more advanced aspects of the problem but the numerical example given below will indicate that a small divergence of the centre of gravity from the centre of rotation may cause quite serious forces.

Numerical Example. *A flywheel weighing 5 tons has its centre of gravity $\frac{1}{10}$ of an inch from the centre of the shaft. Find the force upon the shaft caused by the lack of balance when running at 200 revolutions per minute.*

In this case we may use equation (9), thus getting

$$F_c = \cdot00034 N^2 r W$$
$$= \cdot00034 \times 200 \times 200 \times \tfrac{1}{120} \times 5$$
$$= \underline{\cdot57 \text{ ton.}}$$

If we wish to balance an unbalanced body we may add a weight to it at such a point that it will cause a centrifugal force

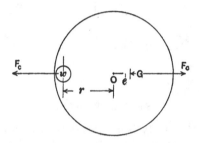

Fig. 153.

equal and opposite to that caused by the eccentricity. Suppose for instance that a body of weight W is rotating about an axis O, Fig. 153, and that its centre of gravity G is at a distance e from O.

This causes a centrifugal force F_c which may be balanced by an equal and opposite force F_c caused by a weight w placed at radius r at a point diametrically opposite to G.

Now this means that we shall bring the centre of gravity of the whole body including the weight w back to O.

Therefore by moments about O we shall have

$$wr = We,$$

$$\therefore \quad w = \frac{We}{r} \quad \dots\dots\dots\dots\dots\dots(13)$$

We can get the same result by considering the centrifugal force F_c.

We then have
$$F_c = \cdot00034N^2eW$$
$$= \cdot00034N^2rw,$$

i.e.
$$rw = eW,$$

or
$$w = \frac{We}{r} \text{ as before.}$$

Projectiles. If a body such as a stone or a bullet is projected into the air in a direction other than vertical it describes a curved path called the *trajectory* which, as we shall show later, is a parabola if the resistance of the air is not taken into account.

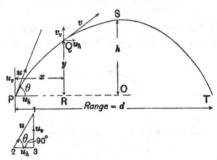

Fig. 154. Projectiles.

Suppose that a body is projected with a velocity u from a point P, Fig. 154, at an inclination θ with the horizontal. Then this velocity may, as we have seen before, be resolved into a vertical component u_v and a horizontal component u_h whose values can easily be found by drawing the $\triangle 123$ to scale or can be found by trigonometrical calculation as follows:

$$u_v = u \sin \theta \ \dots\dots\dots\dots\dots (1),$$
$$u_h = u \cos \theta \ \dots\dots\dots\dots\dots (2).$$

Now the only force acting upon the projectile, if air resistance is neglected, is that of gravity which acts vertically downwards and will give a vertical downward acceleration to the body. But this vertical force can have no effect upon the horizontal component of the velocity, so that *the horizontal component u_h of the velocity remains constant.*

Now suppose that after a time t the projectile has reached a position Q, the components of its velocity v then being u_v and u_h.

Then the vertical distance y will be the same as would be obtained by projecting a body vertically with velocity u_v so that by formula (6) on p. 96 we have

$$y = u_v t - \tfrac{1}{2}gt^2 \dots\dots\dots\dots(3),$$

$$x = u_h t \dots\dots\dots\dots(4).$$

Putting $t = \dfrac{x}{u_h}$, u_h being a constant, we have

$$y = \frac{u_v}{u_h}.x - \tfrac{1}{2}g\frac{x^2}{u_h{}^2}$$

$$= \frac{u_v}{u_h}.x - \frac{g}{2u_h{}^2}.x^2 \dots\dots\dots(5),$$

but this is of the general mathematical form

$$y = ax + bx^2,$$

where a and b are constants, and we know that the corresponding curve is a parabola.

Therefore we have proved that the path of the projectile is a parabola.

We now wish to obtain some way of finding the height h to which the projectile will rise and the range d, i.e. the distance away from P at which the body will again be at the same level.

The height h will be the height to which a body will rise when projected vertically upwards with a velocity u_v.

From equation (12), p. 97, we get

$$h = \frac{u_v{}^2}{2g} \dots\dots\dots\dots(6).$$

The time to reach this point is given by

$$t = \frac{u_v}{g} \dots\dots\dots\dots(7).$$

By the time the body has arrived at T it has been moving for a time $2t$ and since it has been moving in a horizontal direction with a constant velocity u_h, the range d must be given by

$$d = \text{constant velocity} \times \text{time}$$

$$= u_h \times 2t$$

$$= \frac{2u_v.u_h}{g} \dots\dots\dots\dots(8).$$

When we are calculating with the aid of trigonometry we can make results of equations (1) and (2) as follows:

$$d = \frac{2u \sin \theta \,.\, u \cos \theta}{g}$$

$$= \frac{u^2}{g} \,.\, 2 \sin \theta \cos \theta$$

$$= \frac{u^2}{g} \,.\, \sin 2\theta \dots\dots\dots\dots\dots(9).$$

Direction to give greatest range for a given velocity. If the velocity is fixed the range for different directions varies as $\sin 2\theta$ as given by equation (9). Now the greatest value of the sine of an angle is 1 and occurs when that angle is 90°.

Therefore the maximum range occurs when $2\theta = 90°$ or $\theta = 45°$, so that to send a projectile the farthest distance horizontally we should project it at an angle of 45° to the horizon.

Projectiles considered from the hodograph. With a projectile, the acceleration is, as we have seen, constant and is vertically downwards and the horizontal velocity is constant so that in equal times the body moves through equal horizontal distances.

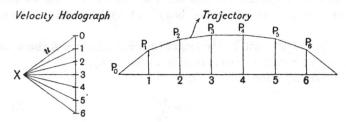

Fig. 155.   Hodograph for Projectiles.

We saw on p. 243 that the acceleration is the velocity with which the point moves on the velocity hodograph so that as this is constant the points on the velocity hodograph are at equal distances apart. This gives us the velocity hodograph shown in Fig. 155. Working backwards from this and the knowledge that in equal times the horizontal distances are equal, we draw a number of vertical lines at equal distances apart and draw P_0P_1 parallel to $X0$, then P_1P_2 parallel to $X1$ and so on. This will be recognised as the link and vector polygon construction which gives a parabola when the points are near enough together.

Consider for instance the graphical construction for the B.M. diagram of a beam carrying a uniform load.

Numerical Examples. (1) *A shot is projected horizontally from the top of a tower 50 feet high with a velocity of 200 feet per second. After what time will it strike the ground and how far away from the base of the tower will it then be?*

In this case the trajectory will be somewhat of the form shown in Fig. 156 and the time taken will be the same as that taken to fall 50 feet from rest.

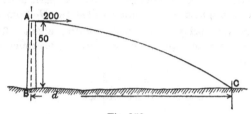

Fig. 156.

Therefore we have from equation (9), p. 97,

$$s = \tfrac{1}{2}gt^2,$$

$$\therefore \quad t^2 = \frac{2s}{g} = \frac{100}{32 \cdot 2} = 3 \cdot 11,$$

$$t = \sqrt{3 \cdot 11} = 1 \cdot 76 \text{ seconds},$$

$$\therefore \quad d = 200 \times 1 \cdot 76 = \underline{352 \text{ feet.}}$$

(2) *If a man can throw a stone 90 yards, how long is it in the air, and to what height will it rise?*

In this case
$$d = 90 \text{ yards}$$
$$= 270 \text{ feet.}$$

And we have seen that $\theta = 45°$ for maximum range so that $u_v = u_h$.

$$\therefore \text{ since} \qquad d = \frac{2u_v \cdot u_h}{g} = \frac{2u_v^2}{g},$$

$$u_v = \sqrt{\frac{gd}{2}}$$

$$= \sqrt{\frac{270 \times 32 \cdot 2}{2}}$$

$$= 65 \cdot 8 \text{ feet per second};$$

∴ from (6) $h = \dfrac{u_v{}^2}{2g} = \dfrac{d}{4} = 67\cdot5$ feet,

from (7)   Time to top $= \dfrac{u_v}{g}$,

∴ Total time $= \dfrac{2u_v}{g} = \dfrac{65\cdot8}{16\cdot1}$

$= \underline{4\cdot1 \text{ seconds nearly.}}$

SUMMARY OF CHAPTER XV.

The motion of bodies moving in a curved path is conveniently studied by a graphical construction called the *Hodograph*.

The angular velocity (ω) of a body rotating about a fixed axis is equal to the angle through which the body rotates round the axis in a unit of time.

$$\therefore \quad v = \omega r = 2\pi rN.$$

Centripetal acceleration $= \dfrac{v^2}{r}$.

Centripetal or centrifugal force $= \dfrac{Wv^2}{gr}$

$$= \cdot00034N^2rW.$$

Governors :

$$h = \dfrac{gr^2}{v^2},$$

$$N = \dfrac{60}{2\pi}\sqrt{\dfrac{g}{h}}.$$

Projectiles. The path of a projectile is called its *trajectory* and if air resistance is neglected it will be a parabola. The horizontal component of the velocity of a projectile is constant.

The vertical component has gravity acceleration acting against it.

Range of projectile $= d = \dfrac{u^2}{g}\sin 2\theta$.

The greatest range on a horizontal plane for a given initial velocity occurs when the angle of projection is 45°.

EXERCISES. XV.

1. A body weighing 2 tons moves in a circle of radius 10 ft. 6 ins. making 180 revolutions per minute. Find its kinetic energy in ft.-lbs.

2. A weight of 1 lb. is fastened to the end of a string 3 ft. long and made to perform 50 revolutions per min. with uniform velocity, the revolutions taking place in a horizontal plane.

Determine the tension of the string.

3. Find the speed at which a simple Watt governor runs when the arm makes an angle of 38° with its vertical. Length of arm from centre of pin to centre of ball = 18 inches.

4. A railway carriage of weight 2 tons is moving at the rate of 60 miles per hour on a curve of 770 ft. radius. If the outer rail is not raised above the inner, find the lateral pressure on the rail.

5. A string 4 ft. long which can just support a weight of 9 lbs. without breaking is placed on a horizontal table. To one end is fixed a weight of 8 lbs. and the free end is held and the weight is swung round. Find how fast the weight may go so as just not to break the string.

6. At what speed must a locomotive be running on level lines with a curve of 968 ft. radius if the thrust on the rails is $\frac{1}{64}$ of its weight?

7. A locomotive engine weighs 38 tons and travels round a curve of 800 ft. radius at 50 miles per hour. Find the centrifugal force. Show how to find the direction and magnitude of the resultant thrust on the rails due to its weight and the centrifugal force.

8. A motor car moves at constant speed in a horizontal circle 300 ft. radius. The track is at 10° to the horizontal. The plumbline makes 12° with what would be perpendicular if the car were on the horizontal. Find the speed of the car.

9. A flywheel 5 ft. 3 ins. in diameter has a rim weighing 1000 lbs. Find the number of foot-pounds of work required to set this rotating 120 times per minute.

10. A brake wheel 4 ft. in diam. on a horizontal axle is furnished with internal flanges which, along with the rim, form a trough containing cooling water. What is the least speed which will prevent the water from falling out?

11. Find the greatest range which a projectile with an initial velocity of 1600 ft. per sec. can attain on a horizontal plane.

12. A rifle has a range of 1000 yards. What would the range be under the same circumstances if fired in the moon where the force of gravity is $\frac{1}{6}$ that of the earth?

CHAPTER XVI

MECHANISMS

FOR our present purpose we will regard a "mechanism" as a device for transferring motion from one point to another in a machine. In many cases the kind of motion becomes changed in the transformation, for instance a rotation becomes changed into an oscillation or a reciprocation or vice versa. The name "linkage mechanism" is used for those mechanisms in which rods are employed which are pivoted together, such rods being called links or elements, and the whole collection of rods being called a "kinematic chain."

Crank and Connecting-rod or Steam-Engine Mechanism. This is about the most common linkage mechanism employed in machinery, and it is used for converting a reciprocating motion

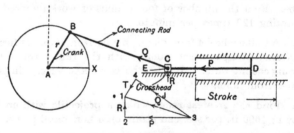

Fig. 157. Crank and Connecting-rod Mechanism.

into a rotary motion or vice versa. It is used on nearly all steam, oil or petrol engines, in which the reciprocation of the piston is converted into a rotation of the shaft, and in a very large number of mechanical presses in which it is employed to convert the rotary motion of a shaft into the reciprocating motion of a press-head.

The mechanism consists of a link AB, Fig. 157, called the *crank*, which is fixed to a rotating shaft and is pivoted at its end B to a rod BC called the *connecting-rod*. The connecting-rod is pivoted at its other end C to a block E called the *cross-head* which is guided so as to move in a straight line and is connected by a piston-rod to the piston D of the engine. On the rotation of the crank the cross-head is caused by its guides to reciprocate. It is interesting to note that James Watt did not use this mechanism for his steam-engine because one of his workmen had stolen the idea and obtained prior patent rights for it. He devised what is called "the Sun and Planet mechanism" which is practically never used nowadays, and the consideration of which is outside our scope.

In this and all other mechanisms to be described the student *must* trace out the movement by actually drawing the mechanism to scale in a number of its possible positions, or else by making a model of the mechanism and attaching a pencil to the point whose motion he wishes to study. The pencil will then trace out on a piece of paper the path in which that particular point moves. Such models can be very easily made by the aid of the constructional toys now on the market.

Velocities in Mechanism. Instantaneous or Virtual Centre. Suppose that a body as shown shaded in Fig. 158 is moving

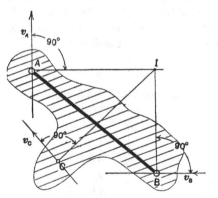

Fig. 158.   Instantaneous or Virtual Centre.

in any manner and the velocities v_A, v_B of two points A and B in it are known in magnitude and direction. Draw AI perpendicular to v_A, and BI perpendicular to v_B, then the intersection

17—2

I is called the *instantaneous or virtual centre* because A and B may both be regarded as rotating for the instant about this point. We may therefore study the motion of the body at the particular instant under consideration by imagining it to be rotating about the point I. It is important to remember that unless the body is rotating about a fixed point, I will be constantly changing and the curve in which I moves is called the "centrode." At any instant, however, we can find the relation between the velocities of the various points of the body if we know the instantaneous centre, because when a body is rotating we have seen that the velocity of any point in it is proportional to the radius of the point. We therefore have

$$\frac{v_A}{v_B} = \frac{AI}{BI}.$$

To obtain the velocity of any other point, say C, we join CI and draw a line at right angles to it. This gives the direction of v_C, and its magnitude is given by the relation

$$\frac{v_C}{v_A} = \frac{CI}{AI}.$$

Application to Crank and Connecting-rod Mechanism. Suppose that the shaft A, Fig. 159, is rotating uniformly so that the crank pin B has a uniform velocity v_B, at right angles to the crank. Draw CI perpendicular to the direction of the cross-head and produce AB to meet it at I. Then I will be the instantaneous centre of the movement of the connecting-rod BC. Because BI is at right angles to v_B, we therefore have

$$\frac{v_C}{v_B} = \frac{CI}{BI} \quad \dots\dots\dots\dots\dots\dots\dots(1),$$

but since the triangles BIC and BAD are similar we have

$$\frac{CI}{BI} = \frac{AD}{AB},$$

$$\therefore \frac{v_C}{v_B} = \frac{AD}{AB},$$

that is to say $$v_C = \frac{v_B}{AB}.AD \quad \dots\dots\dots\dots\dots(2).$$

If therefore we choose our scale of velocity so that AB represents v_B the length of the crank pin AD will give us the velocity of the cross-head and therefore of the piston to the same scale. By repeating this construction for a large number of positions of the crank pin and cross-head we can find the velocities in the different positions, and from these we can draw a diagram showing the manner in which the velocity varies. Two convenient forms of diagrams are shown in the figure. One is drawn upon a base of the stroke and is obtained by projecting the point D upon the line CI, thus obtaining the point F, and joining up points such as F. This diagram is useful when we wish to find the velocity for a given position of the cross-head or piston. The

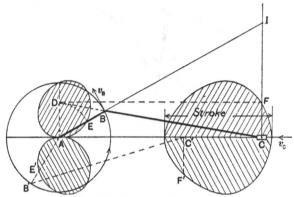

Fig. 159.   Velocity Diagrams.

other form shown is called a *polar diagram* and is extremely useful for finding the velocity of the piston for different positions of the crank. It is obtained by drawing with centre A an arc of radius AD to meet the crank (produced if necessary) at E and joining up the points thus obtained. It will be found to give two loops as shown. In the use of these diagrams for any position of the cross-head, say C', the velocity of the cross-head is given by $C'F'$ or by AE'.

Force in connecting-rod; crank effort. The force Q in the connecting rod can be found by drawing the triangle of forces 1, 2, 3 as indicated in Fig. 157. This force Q can be resolved into a component T along a component J at right angles to the crank. The force J is called the "crank effort." If no work is lost, the work done by P per second must be equal to that done by J,

$$\therefore J \cdot v_B = P \cdot v_C,$$

or
$$J = \frac{P \cdot v_C}{v_B}$$
$$= \frac{P \cdot AD}{AB} \text{ (Fig. 159)} \quad \ldots \ldots (3).$$

Watt's Parallel Motion. This mechanism was used by James
Watt to guide the valve rods of his beam engines without the
necessity of providing a cross-head and was regarded by him as
one of the most ingenious of his inventions. The rod AB,
Fig. 160, is pivoted at the point A and is connected by a "coupler"

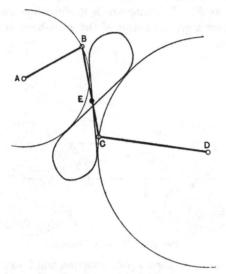

Fig. 160.   Watt's Parallel Motion.

BC to a rod CD pivoted at the opposite side as shown.   A point
E is taken on BC such that

$$\frac{BE}{EC} = \frac{CD}{AB},$$

and as one or other of the rods AB or CD is oscillated the point
E will be found to move in a line which is for all practical purposes
straight for small amounts of oscillation. For a complete
revolution of AB the point E will be found to trace out a looped
figure as shown.

Slotted lever quick-return mechanism. This mechanism is
used to reciprocate the ram which carries the cutting tool of
shaping machines and it has the property that the time taken in
the forward or cutting movement of the ram is greater than the
time taken in the return or idle movement; this is economical

because it reduces the time during which the cutting tool is doing no useful work. This mechanism is often called the "Whitworth quick-return motion," but this description is not quite correct, Whitworth's mechanism being slightly different although possessing the same property.

The mechanism consists of a lever AB (Fig. 161) pivoted at the lower end to a fixed point A and provided with a slot C in which works a crank pin D which rotates about a centre E vertically above the fixed point A. The upper end B of the

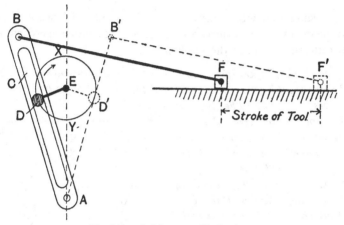

Fig. 161. Quick-return Mechanism.

slotted lever is connected by a rod BF which is pivoted at the end F to the ram of the shaping machine. This ram runs in horizontal guides and it is usual to provide means for adjusting the position of the crank pin D so as to alter the stroke of the ram. The mechanism is shown diagrammatically in the figure in its extreme position. While the crank pin is moving through the arc DXD' the tool ram moves from F to F' thus effecting the cutting stroke and while the pin moves through the arc $D'YD$ the ram makes the idle or return stroke. If the crank pin rotates with uniform velocity, the time taken on the cutting stroke must be proportional to the length of arc DXD' and on the return stroke to the arc $D'YD$; and since the arcs are proportional to the angles that they subtend at the centre of the circle and therefore are also proportional to the halves of such angles, the cutting and return times are respectively proportional to the angles *DEX* and *DEY*.

$$\therefore \text{ we have } \quad \frac{\text{cutting time}}{\text{return time}} = \frac{\angle DEX}{\angle DEY}.$$

But since the mean cutting speeds are inversely proportional to the corresponding times we have

$$\frac{\text{mean return speed}}{\text{mean cutting speed}} = \frac{\text{cutting time}}{\text{return time}}$$

$$= \frac{\angle DEX}{\angle DEY}.$$

It should be noted that this ratio deals only with the average or mean cutting or return speeds, because the actual speed varies at different points of the stroke.

Numerical Example. *In a Whitworth quick-return gear of a shaping machine the stroke is 8" and the ratio of home and cutting strokes is* 3 : 5. *The line of stroke of the ram produced passes through the extreme positions of the connecting-rod pin at the end of the slotted lever. If the distance between the centre of the driving plate and the axis about which the slotted lever oscillates is 6", find the crank radius and length of the lever.*

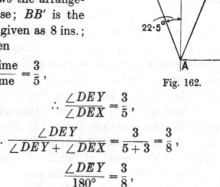

Fig. 162.

Fig. 162 shows the arrangement in this case; *BB'* is the stroke which is given as 8 ins.; we are also given

$$\frac{\text{cutting time}}{\text{return time}} = \frac{3}{5},$$

$$\therefore \frac{\angle DEY}{\angle DEX} = \frac{3}{5},$$

$$\therefore \frac{\angle DEY}{\angle DEY + \angle DEX} = \frac{3}{5+3} = \frac{3}{8},$$

that is

$$\frac{\angle DEY}{180°} = \frac{3}{8},$$

or

$$\angle DEY = \tfrac{3}{8} \times 180° = 67·5°.$$

Now $\angle DAE = 90° - \angle DEY = 22\cdot5°$,

and $\angle ABK = 90° - 22\cdot5° = 67\cdot5.°$

We are now in a position to draw the figure to scale, and by first drawing BB' horizontal to represent 8 ins. and then $\angle B'BA$ and $\angle BB'A$ each $= 67\cdot5°$ we get the point A and AB which is the required length of the lever and will be found by measurement to be about 10·5 ins.

Now set up $AE = 6$ ins. and draw ED perpendicular to AB.

Then ED is the crank radius and will be found to be about 2·29 ins.

If as is preferable we proceed by trigonometrical calculations we shall have

$$\frac{BK}{AB} = \sin 22\cdot5°,$$

$$\therefore AB = \frac{4}{22\cdot5} = 10\cdot46 \text{ ins.}$$

Also $$\frac{ED}{EA} = \sin 22\cdot5°,$$

$$\therefore ED = EA \sin 22\cdot5° = 6 \sin 22\cdot5° = 2\cdot29 \text{ ins.}$$

Toggle Mechanism. The name "toggle" is used to denote a linkage mechanism in which one part receives a very small

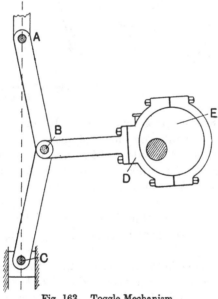

Fig. 163.   Toggle Mechanism.

motion while another receives an appreciable movement; it is
used as a means of exerting heavy pressures in presses and is
also used in a large number of every-day appliances, such as the
devices which are to be found for closing bottles of various kinds.
Fig. 163 shows a common arrangement for use in mechanical
presses. The toggle links AB, BC are connected at one end A to
a fixed support and at the other end C to the press-head. The
joint B is connected to an eccentric D carried by a rotating shaft
E the arrangement being such that as the shaft rotates the
eccentric is reciprocated and a small movement is given to the

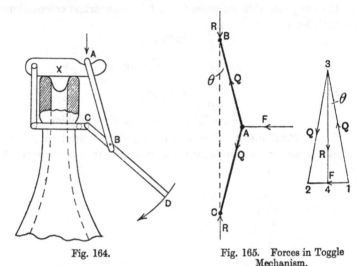

Fig. 164.   Fig. 165.   Forces in Toggle
                                         Mechanism.

press-head which exerts a very considerable pressure. Fig. 164
shows diagrammatically one form of toggle closing device for
stoppered bottles. A wire loop AB passes over a groove on top
of the stopper X and is pivotally connected at the point B to
a bent wire lever CD which is pivoted at the point C to a wire
ring fastened round the neck of the bottle. As the point D is
moved about the centre C in the direction of the arrow the loop
AB is pulled downwards and exerts a strong closing action upon
the stopper.

We can examine in the following manner the pressure exerted
in the toggle press for any position of the toggle levers. Referring
to Fig. 165 the force or effort F exerted by the eccentric is resolved

into two forces Q acting down the toggle links, and these forces
Q can be resolved into vertical components R one of which at
the point B is carried by the framing of the machine and the
other of which is the pressure exerted on the press-head at C.
The $\triangle\ 123$ is the triangle of forces and from this we get that

$$\frac{R}{\dfrac{F}{2}} = \frac{3,4}{4,1}$$

$$= \cot \theta,$$

i.e. $$R = \frac{F}{2} \cot \theta.$$

A glance at the trigonometrical tables will show that as
an angle gets small its cotangent increases very rapidly so
that we see that if the angle θ is small the pressure R will be
very many times more than the force F. In the limiting condition $\theta = 0$, the pressure would theoretically be infinitely great,
but in the practical use of the mechanism θ can never be exactly
zero although it may be very near to it, as in practice there is a
limit to the pressure which the mechanism can exert owing to
the yielding of the various parts composing it.

Cams and Wipers. Cams, or wipers as they are sometimes
called, are a form of mechanism for converting rotary motion
into a reciprocating or oscillating motion.

Fig. 166 shows a cam C for giving a reciprocating motion to
a shaft E from a rotating shaft A. The shaft E is guided by
a slide G and carries at its end an anti-friction roller D which
rides upon the face of the cam. A spring F is employed for

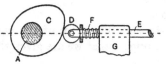

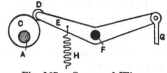

Fig. 166.   Cams and Wipers.   Fig. 167.   Cams and Wipers.

keeping the roller in contact with the cam when the latter is in
such a position that the shaft E is moving towards the shaft A.
Other devices for this purpose are sometimes devised, a common
one being that the roller runs in a groove cut in the cam disc
and is thus positively moved in both directions. The form
shown in the figures is, however, usually preferable.

In Fig. 167 is shown a cam C communicating an oscillating

motion to a lever E pivoted at F. A rod G pivoted at the other end of the lever communicates the motion to the required part of the machine, and a spring H keeps the roller D in contact with the cam. Cams of this kind are used in almost every form of gas-engine for operating the valves in the required sequence. The cams shown in Figs. 166, 167 are often called *plane or edge cams*, the form shown in Fig. 168 being called a *surface or drum cam*. In the latter case the cam is formed as a groove in a drum C carried by a rotating shaft A. A roller carried by a lever D engages the groove and the lever is pivoted at E and connected at its extreme end to a slide F to which the cam communicates the required motion. The lever is stationary while the centre of the groove remains in a plane section of the drum normal to the axis. In the case of plate cams the slide or lever is stationary for the portion of the cam that is concentric with the axis of the shaft.

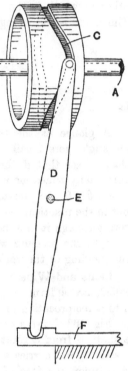

Fig. 168.   Surface or Drum Cam.

Design of a Plate Cam. A plate cam for giving a reciprocating movement in a straight line passing through the centre of the cam shaft can be designed as follows. Suppose that we are given the following particulars:

The diameter D of the cam shaft.

The diameter d of the roller.

The minimum thickness t of the cam.

The lift or height h of the movement of the roller.

The manner in which the roller has to rise and fall; this is called the "timing" of the cam.

To make our illustration more clear we will assume that our cam is required to move the slide uniformly upward during

one-third of a revolution, is then required to remain stationary
for another third of a revolution and has finally to fall uniformly during the remainder of the revolution.

First draw with centre O a circle of diameter D (Fig. 169) to
represent the cam shaft, and draw the line OX along which the
roller has to reciprocate. Next make CD equal to t, the minimum
thickness of the cam and find the centre A of the roller in its lowest
position. AB is next set up equal to the lift h and AB is divided

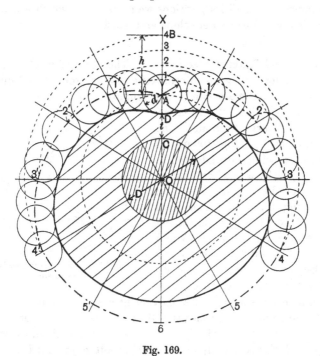

Fig. 169.

up into a convenient number of equal parts, say 4. A number of
equally spaced radial lines $O1$, $O2$, etc. are then drawn, 12 being
taken as a convenient and sufficient number.

In moving through one-third of a turn, i.e. from OA to $O4$,
we have to rise a height h and have to do so uniformly; we
therefore make $O1 = O1$ and $O2 = O2$; $O3 = O3$ and $O4 = O4$ as
indicated by the circular arcs shown in dotted lines. During the
next one-third of a revolution the roller has to remain stationary
so that we draw an arc with centre O from 4 to 4′, and since the

roller has to fall uniformly during the remainder of the revolution we repeat the construction for the points 1, 2, 3, 4 to get the points 1', 2', 3', 4'.

By joining up the points thus obtained we obtain the curve shown in chain dotted lines, this curve being that for the centre of the roller. We then go round the curve and draw the roller circle all round it to give the effect of the roller running along such a curved path. By drawing a line to touch the roller on the inner side in all its positions we get a curve which a mathematician would call an envelope and which gives us the shape of the cam required.

Pawl and Ratchet Mechanism. This form of mechanism is employed for giving an intermittent motion from a continuous motion and is most commonly employed for converting an oscillating motion into an intermittent rotary motion. The ratchet wheel A, Fig. 169 a, has teeth on it which are adapted to be engaged by a pawl or "click" B carried by a pivot D on an oscillating lever C. The pawl drives when the lever is moved in the direction of the arrow and slips over the ratchet when moved in the opposite direction and re-engages one of the teeth beyond,

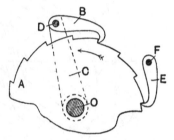

Fig. 169 a.   Pawl and Ratchet Mechanism.

when it reverses again. The movement given to the pawl should be a whole number of times the distance apart of the teeth. To prevent the ratchet wheel from moving back with the pawl, a stop-pawl E, pivoted upon a pin F, is often provided.

The ratchet brace gives a very familiar example of this mechanism.

SUMMARY OF CHAPTER XVI.

The crank and connecting-rod mechanism is a device for convert-ing a reciprocating motion into a rotary motion and vice versa.

The *virtual centre* of a body moving in any manner is a point about which the body may be regarded as rotating at any particular instant.

The path moved through by the virtual centre is called the *centrode*.

Watt's parallel motion. The point which has to move in a straight line divides the coupler inversely as the lengths of the pivoted rods.

Slotted lever mechanism. The return or idle motion is quicker than the forward or cutting motion, the relative values are obtained by considering the angles turned through by the crank pin in the extreme positions.

Toggle mechanism is employed to obtain a heavy pressure moving through a small distance from a small force moving through a greater distance. The forces are obtained by the ordinary triangle of forces.

Cams are a device for giving an oscillating or reciprocating motion from a constantly rotating shaft. They may be " plane or edge " cams or " surface or drum " cams.

Ratchets are a form of mechanism for obtaining an intermittent motion from a constant reciprocating or rotary motion.

EXERCISES. XVI.

1. The connecting-rod of an engine is $2\frac{1}{2}$ times the stroke in length. Find graphically (a) the position of the crank when the piston is at half-stroke, (b) the position of the piston when the crank is 90° from a dead point.

2. The stroke of an engine is 2 ft. and connecting-rod 4 ft. long. The thrust on the piston is 12,000 lbs. When the crank is vertical find

      (1) The thrust on the cross-head.

      (2) The thrust on the connecting-rod.

3. The connecting-rod of a steam-engine is 4·5 ft. long and the crank has a radius of 1·5 feet. Draw a curve showing the displacement of the piston for different angular positions of the crank.

4. In a steam-engine mechanism the crank radius is 10 inches and the connecting-rod 10 inches long. If the crank makes 120 revs. per min. find the velocity of the piston when the connecting-rod is at right angles to the crank.

5. Trace the curve drawn by the Watt parallel motion if the length of the pivoted links is 24″ and the coupling-rod 12″, the upper pivoted link having an oscillation 20° from the horizontal.

6. In a quick-return motion for a shaping machine the length of the lever is 10·46 ins. and the crank radius 1·15 ins., the distance between the centre of the rotating shaft and the pivot of the slotted lever being 6 ins. Find the ratio of return to cutting speeds and the stroke of the cutting tool.

CHAPTER XVII

BELT, CHAIN, AND TOOTHED GEARING

BELT, chain and toothed gearing is a form of mechanism for converting a rotary motion about a certain centre into a rotary motion about another centre. In the case of a belt, the power is transmitted through the friction between the belt and the pulleys, and in the case of toothed gearing the power is transmitted through the stresses in the material of the teeth. Chain gearing is similar to belt gearing except that in place of the friction drive we have positive drive between the teeth of the sprocket wheels and the links of the chain.

Belt Gearing. If the belt transmits motion from a shaft X to a shaft Y the pulley on the shaft X is called the *driver* and that on the shaft Y is called the *follower*. The belts may be *open* as (a) (Fig. 170) or *crossed* as (b). In the open arrangement the driver and follower rotate in the same direction, whereas in the crossed arrangement they rotate in opposite directions. The power that can be transmitted by the gear depends upon the friction between the belt and the pulley, and this friction depends on the angle subtended at the centre of the pulley by the arc of contact.

In the crossed arrangement this angle is greater than in the open arrangement so that in this respect the crossed arrangement is better than the other. The tension T_1 in the belt on the side as it comes on to the driver is greater than the tension T_2 on the other side. If the driver is D_X feet in diameter and makes N_X revolutions per minute and the follower is D_Y feet in diameter and makes N_Y revolutions per minute, the work done against the tension T_1 on the tight side per minute will be equal to the force multiplied by the distance moved by the belt per minute

$$= T_1 . \pi D_X N_X.$$

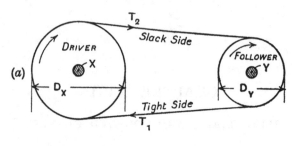

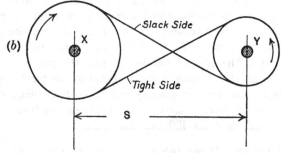

Fig. 170.   Belt Gearing.

Similarly the work done on the pulley by the belt on the slack side

$$= T_2 . \pi D_X N_X,$$

∴ total work done per minute on belt

$$= T_1 . \pi D_X N_X - T_2 . \pi D_X N_X$$
$$= (T_1 - T_2) . \pi D_X N_X \text{ ft.-lbs.},$$

∴ H.P. transmitted $= \dfrac{\text{work done per minute in ft.-lbs.}}{33{,}000}$

$$= \frac{\pi D_X N_X (T_1 - T_2)}{33{,}000} \quad \dots\dots\dots\dots(1).$$

In the absence of other information T_1 may be taken as twice T_2.

When the diameter of the belt is appreciable compared with the diameter of the pulley for calculations D_X should be measured to the centre of the belt.

Numerical Example. *Find the* H.P. *that can be transmitted by a pulley* 3 *ft.* 6 *ins. diameter running at* 120 *revolutions per minute by a thin belt* 6 *ins. wide if the permissible tension in the belt is* 80 *lbs. per in. of width and the tension on the slack side is equal to half that on the tight side.*

In this case $\qquad T_1 = 6 \times 80 = 480$ lbs.,

$$T_2 = 240 \text{ lbs.}$$

$$\therefore \ T_1 - T_2 = 240 \text{ lbs.}$$

$$\therefore \text{ H.P. transmitted} = \frac{\pi \cdot 3 \cdot 5 \times 120 \times 240}{33,000} = \underline{9 \cdot 6}.$$

Velocity Ratio in Belt Gearing. Now by exactly similar reasoning applied to the follower instead of the driver we should get

$$\text{H.P. transmitted} = \frac{\pi \cdot D_Y \cdot N_Y (T_1 - T_2)}{33,000} \quad \dots\dots(2),$$

and if no power is lost these two must be equal.

$$\therefore \ D_X N_X = D_Y N_Y,$$

$$\therefore \text{ velocity ratio} = \frac{\text{no. of revolutions per min. of follower}}{\text{no. of revolutions per min. of driver}}$$

$$= \frac{N_Y}{N_X} = \frac{D_X}{D_Y},$$

$$\therefore \ v_R = \frac{\text{diameter of driver}}{\text{diameter of follower}} \quad \dots\dots\dots\dots(3).$$

We could have obtained this result rather more simply without going into the question of H.P. Unless the belt slips on the pulley, the length of the belt passing on to the driver per minute

= circumference of driver × no. of revolutions per minute

$= \pi D_X N_X.$

But unless the belt stretches this must be exactly equal to the length of the belt passing on to the follower per minute

$$= \pi D_Y N_Y,$$

$$\therefore \ \pi \cdot D_X N_X = \pi \cdot D_Y N_Y,$$

$$\therefore \ D_X N_X = D_Y N_Y.$$

Numerical Example. *A shaft running at* 120 *revolutions per minute carries a belt pulley of* 3 *ft.* 6 *ins. diameter. What must be the diameter of the pulley on the shaft driven by the belt if it runs at* 300 *revolutions per minute?*

In this case $N_X = 120$, $D_X = 3\cdot5$ and $N_Y = 300$,

$$\therefore \text{ since } N_X D_X = N_Y D_Y,$$

$$300 D_Y = 120 \times 3\cdot5,$$

$$\therefore D_Y = \frac{120 \times 3\cdot5}{300}$$

$$= 1\cdot4 \text{ feet.}$$

Belt Speed-cones. In belt-driven machines it is often desirable to vary the velocity ratio transmitted, i.e. to vary the speed of the machine. This is usually effected by speed-cones which consist of two sets of pulleys whose sizes are so arranged that the belt will run tightly between any opposite pair.

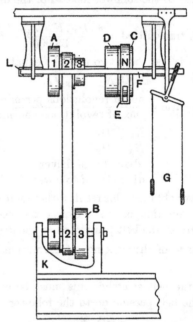

Fig. 171.   Belt Drive for Lathe Headstock.

Fig. 171 shows an arrangement commonly employed for driving a lathe. A cone of three pulleys B is mounted in the headstock K of the lathe and an overhead shaft L carries a corresponding cone of pulleys A. When the belt is between the pulleys 1, the driver is larger than the follower and we then have the quickest speed of the headstock spindle; when the belt is in the position 2 shown, the two pulleys are about equal in diameter so that the speed is less; whereas when the belt is between the pulleys 3 the driver is smaller than the follower so that the headstock spindle is driven at its lowest speed.

Belt-striking gear. Fig. 171 shows also one form of device used for starting and stopping a machine driven by belt gearing. The overhead shaft is driven by a belt N and two pulleys C, D of equal diameter are placed alongside on the shaft. The pulley C is keyed to the shaft, and is called the "fast pulley," and the pulley D is loosely mounted and is called the "idle or loose pulley." The belt N passes between forks E carried by a sliding rod F which is moved lengthwise by a slotted lever which is moved to one or other of its extreme positions by means of chains G.

In the position shown, the belt N is upon the fast pulley C so that the headstock is being driven. If the left handle G be pulled down the rod F will be moved to the left and the belt N will be moved on to the idle pulley D which does not drive the shaft L because it is not keyed to it.

Sizes of Cones for keeping Belt taut. As we have already indicated, it is necessary that the diameter of the pulleys in a cone shall be such that the same length of belt will run taut over all of them.

Open belts. If S is the distance apart of the shafts, the length of an open belt is given approximately by the formula

$$l = 2S + \frac{\pi}{2}(D_X + D_Y) + \frac{(D_X - D_Y)^2}{4S} \quad \ldots\ldots(4).$$

If therefore we are given S, and the diameters D_X and D_Y of one pair of pulleys, the diameters of the others should be chosen so as to keep l practically constant.

Crossed belts. In this case it can be proved that the length of belt is constant for a fixed value of S if the sum of the diameters of the pulleys is constant so that it is quite an easy matter to choose suitable diameters of a cone of pulleys to work with crossed belts.

Belt Reversing Gear. In some machines, such as planing machines, it is necessary to reverse periodically the direction of rotation of the working parts. With a belt drive this can be

effected by the arrangement shown in plan in Fig. 172 which
we will describe with reference to a planing machine.

The main driving shaft B carries a broad pulley A upon which
are carried a crossed and an open belt. The driven shaft D
carries two outside idle pulleys and a central fast pulley. Belt forks
C are provided and are so spaced that one belt is on the fast
pulley and one on an idle pulley. In the position shown the
open belt is driving and the cross belt is idly rotating its pulley
in the reverse direction. When the planing machine later reaches
the end of its stroke, tappets or blocks adjustably mounted upon
it strike arms which communicate their motion to the shaft
carrying the belt forks C. The latter are then moved (upward
on the drawing) so that the crossed belt comes on the fast pulley

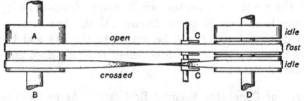

Fig. 172.   Belt Drive for a Planing Machine.

and the open belt moves on to the upper idle pulley. The
directions of the rotation of the shaft D and therefore of the
movement of the machine table are thus reversed until the table
reaches the other end on its stroke whereupon the parts are
returned to the position shown and the cycle of operations is
repeated.

The fast pulley is often made narrower than the idle pulleys
and the distance apart of the belt forks arranged so that in an
intermediate position of the latter each belt is on an idle pulley;
in this way the operation of the machine can be stopped when
required.

Belt Drive for Inclined Axes. Up to the present we have
considered only the case in which the axes of the shafts to be
driven by belting are parallel.

Fig. 173 shows a way of providing for two axes at right
angles to each other provided that they are not too close. This
will drive satisfactorily in the direction shown; the arrangement

must be such that the middle point of the width of the belt where it leaves one pulley is in the central plane of the other pulley.

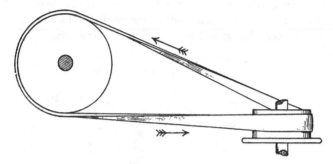

Fig. 173. Belt Drive for Inclined Axes.

Where the position of the shafts is such that a direct drive cannot be effected, guide pulleys must be used. Fig. 174 shows one such arrangement, G, G being the guide pulleys.

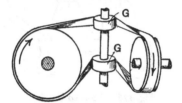

Fig. 174. Belt Drive with Guide Pulleys.

Toothed Gearing. Suppose that two smooth discs X, Y (Fig. 175) rotate in contact without slipping.

Then in one revolution of the driver X a point on the circumference moves through a distance πD_X. If therefore there is no slip between the two discs, a point on the circumference of the follower Y must move through the same distance πD_X. But one revolution of the follower corresponds to πD_Y, so that the number of revolutions of the follower for one of the driver

$$= \frac{\pi D_X}{\pi D_Y} = \frac{D_X}{D_Y},$$

$$\therefore \frac{\text{number of revs. of follower}}{\text{number of revs. of driver}} = \text{velocity ratio} = \frac{D_X}{D_Y}\ldots\ldots(1).$$

Now suppose that in order to prevent any possibility of

slipping we form teeth upon the surfaces of these discs. For simplicity in the figure only a few teeth are shown, but it will be understood that they are formed all round the wheel.

It is clear that these teeth must be of special shape if they are to mesh and roll into action smoothly. The curve most commonly used for gear teeth is called the *involute*.

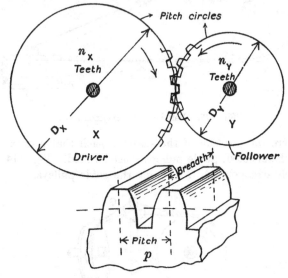

Fig. 175.   Toothed Gearing.

The form of toothed gear shown in Fig. 175 is called *spur gearing*. The axes of the two shafts are parallel and the teeth are straight and usually run at right angles to the plane of the wheels.

The circles which correspond to the untoothed or smooth discs are called the *pitch circles*. The distance upon the pitch circle between the centres of two succeeding teeth is called the *pitch*, or more accurately the *circular pitch p*. It is equal to the circumference of the pitch circle divided by the number of teeth. The *diametral pitch m* is equal to the diameter of the pitch circle divided by the number of teeth and can be obtained by dividing the circular pitch p by π (3·1416). The diametral pitch is sometimes called the *module*. Other forms of toothed gearing are shown in Figs. 176–179.

In the *rack and pinion*, Fig. 176, one of the members is straight, this corresponding to spur gearing in which one of the wheels is

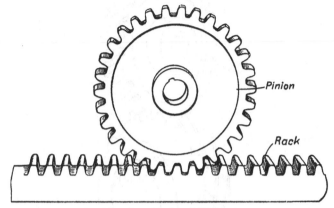

Fig. 176.   Rack and Pinion.

infinitely large. A rotation of the pinion causes a rectilinear movement of the rack.

Bevel gearing, Fig. 177, is used to connect two shafts at an

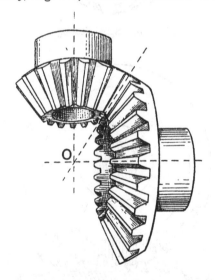

Fig. 177.   Bevel Gearing.

angle to each other (usually a right angle) and meeting at a point. It corresponds to a toothed form of two smooth cones rotating in contact.

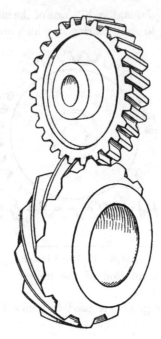

Fig. 178.   Spiral Gearing.

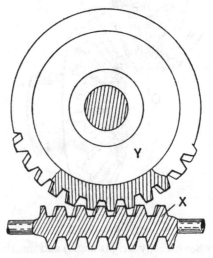

Fig. 179.   Worm Gearing.

Spiral gearing, Fig. 178, is used to connect two shafts at an angle to each other which do not meet at a point.

In *worm gearing*, Fig. 179, the shafts are at right angles to each other and do not intersect. The driver X is a worm or screw and the follower Y is a wheel whose teeth are formed to gear accurately with the worm. The velocity ratio in such gears is small and as a rule the gear cannot be reversed, i.e. the wheel cannot drive the worm. As was shown on p. 66 this means that the efficiency of the gear cannot be greater than 50 %, but in many machines this objection is of minor importance compared with the advantage that the gear is self-locking, i.e. that it will not run backwards if the drive is removed.

Velocity Ratio in Toothed Gear Trains. The term "gear train" is used to indicate a number of gear wheels working in combination.

For a pair of spur wheels we have seen that

$$\text{velocity ratio} = v_r = \frac{\text{diameter of driver}}{\text{diameter of follower}}.$$

Now the circular pitch of the two wheels must be the same and the number of teeth × pitch must be equal to the circumference of the pitch circle.

∴ we have

$$n_X p = \pi D_X \quad \dots\dots\dots\dots\dots (2),$$

and

$$n_Y p = \pi D_Y \quad \dots\dots\dots\dots\dots (3).$$

Dividing we get

$$\frac{n_X}{n_Y} = \frac{D_X}{D_Y} \quad \dots\dots\dots\dots\dots (4).$$

∴ from (1)   velocity ratio $= v_r = \dfrac{n_X}{n_Y}$ $\quad \dots\dots\dots\dots (5).$

Expressed in words:

$$\text{velocity ratio} = \frac{\text{number of teeth on driver}}{\text{number of teeth on follower}},$$

i.e. $\dfrac{\text{number of revs. of follower}}{\text{number of revs. of driver}} = \dfrac{\text{number of teeth on driver}}{\text{number of teeth on follower}},$

$$\text{i.e. } \frac{N_Y}{N_X} = \frac{n_X}{n_Y} = \frac{D_X}{D_Y} \quad \dots\dots\dots\dots (6).$$

Numerical Examples. (1) *A toothed wheel of* 10 *inches diameter on the pitch line and with* 60 *teeth runs at* 120 *revolutions per minute and drives a wheel of* 4 *inches diameter.*

*Find (a) the circular pitch of the teeth, (b) the diametral pitch,
(c) the number of teeth on the second wheel, (d) the number of
revolutions which it will make.*

(a)
$$np = \pi D,$$
$$\therefore\ 60p = 3\cdot1416 \times 10,$$
$$\therefore\ p = \frac{31\cdot416}{60}$$
$$= \cdot524 \text{ inch.}$$

(b)   Diametral pitch $= m = \dfrac{\text{diam. of wheel}}{\text{number of teeth}}$
$$= \tfrac{10}{60}$$
$$= \cdot167 \text{ inch.}$$

(c)   From result (4) we have
$$\frac{n_x}{n_Y} = \frac{D_x}{D_Y},$$
$$\frac{60}{n_Y} = \frac{10}{4},$$
$$\therefore\ 10n_Y = 240,$$
$$\therefore\ n_Y = 24.$$

(d)   In this case $v_r = \tfrac{60}{24}$ from (5)
$$= 2\cdot5.$$

Now $v_r = \dfrac{\text{revolutions of follower}}{\text{revolutions of driver}},$

$\therefore$ revolutions of follower $= 120 \times 2\cdot5 = $ 300 per minute.

(2)   *Toothed wheels of $2\frac{1}{2}$ inches pitch are required to connect two
shafts running at 340 and 115 revolutions per minute, the centres of
the wheels to be as nearly as possible 3 ft. apart. Find suitable
numbers of teeth for the wheels.*

The distance apart of the centres of two toothed wheels is
equal to the sum of the radii of the pitch circles, i.e. equal to
half the sum of the diameters of the pitch circles,

i.e. distance apart $= \dfrac{D_x + D_Y}{2}$

Now we have $\dfrac{N_Y}{N_x} = \dfrac{D_x}{D_Y},$

$\therefore\ \dfrac{D_x}{D_Y} = \dfrac{340}{115} = \dfrac{68}{23}.$

$\therefore\ \dfrac{D_x + D_Y}{D_x} = \dfrac{68 + 23}{23} = \dfrac{91}{23}.$

Now take $D_X + D_Y = 6$ ft. $= 6 \times 12$ inches,

$$\therefore D_X = \frac{6 \times 12 \times 23}{91} = 18\cdot2 \text{ inches,}$$

$$\therefore D_Y = 72 - 18\cdot2 = 53\cdot8 \text{ inches.}$$

Now $n_X p = \pi D_X,$

and $n_Y p = \pi D_Y,$

$$\therefore n_X = \frac{\pi \times 18\cdot2}{2\cdot5} = 22\cdot9,$$

$$n_Y = \frac{\pi \times 53\cdot8}{2\cdot5} = 67\cdot6.$$

But the number of teeth must be a whole number,

$$\therefore \text{ take } n_X = 23 \text{ and } n_Y = 68.$$

These will give the required velocity ratio. The student should note carefully that in problems of this kind it is essential that the numbers of teeth be chosen to give the exact velocity ratio required.

Idle Gear Wheels. In the use of spur gearing it is often necessary to use wheels intermediate between the driver and the follower, as shown in Fig. 180, such wheels being called "idle

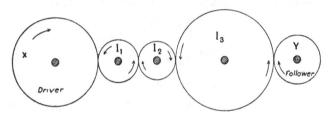

Fig. 180.   Idle Gear Wheels.

wheels" or "idlers." These idle wheels are used either to reverse the direction of rotation or else to enable the distance between the two shafts to be greater than the sum of the radii of the driver and follower.

Idle wheels have no effect on the velocity ratio. If the number of idle wheels is odd the driver and follower rotate in the same direction, but if even they rotate in opposite directions.

Compound Gear Trains. To obtain a larger or smaller velocity ratio than is practicable with one pair of spur wheels, compound gear trains such as shown in Fig. 181 are usually employed. Such a gear train will, for instance, be found in every watch or clock, the form shown giving a small velocity ratio.

The wheel A gears with a wheel B which is formed solid with or is keyed to the same shaft as a wheel C; this drives a spur gear D which is coaxial with a wheel E which gears with the follower F. It is usual to refer to the alternate wheels A, C, E as drivers and the wheels B, D, F as followers.

We will trace out the compound velocity in steps n_A, n_B etc., being the number of teeth in the various wheels, and N_A, N_B etc., their number of revolutions per minute, it being noted that N_B must be equal to N_C and N_D equal to N_E.

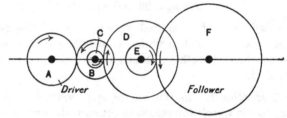

Fig. 181. Compound Gear Train.

Considering the first pair of wheels A, B, we have

$$\frac{N_B}{N_A} = \frac{n_A}{n_B},$$

$$\therefore \; N_B = N_A \cdot \frac{n_A}{n_B} \; \dots \dots \dots \dots \dots (7),$$

$\therefore$ since $N_C = N_B$,

$$N_C = N_A \cdot \frac{n_A}{n_B} \; \dots \dots \dots \dots \dots (8).$$

Considering the wheels C and D, we have

$$\frac{N_D}{N_C} = \frac{n_C}{n_D},$$

$$\therefore \; N_D = N_C \cdot \frac{n_C}{n_D}$$

$$= N_A \cdot \frac{n_A}{n_B} \cdot \frac{n_C}{n_D} \; \text{(from 8)} \; \dots (9),$$

$\therefore$ since $N_E = N_D$,

$$N_E = N_A \cdot \frac{n_A}{n_B} \cdot \frac{n_C}{n_D} \; \dots \dots \dots \dots (10).$$

Considering the wheels E and F, we have

$$\frac{N_F}{N_E} = \frac{n_E}{n_F},$$

$$\therefore\ N_F = N_E \cdot \frac{n_E}{n_F}$$

$$= N_A \cdot \frac{n_A}{n_B} \cdot \frac{n_C}{n_D} \cdot \frac{n_E}{n_F} \dots\dots (11).$$

$$\therefore\ \frac{N_F}{N_A} = \text{velocity of compound train}$$

$$= \frac{n_A \times n_C \times n_E}{n_B \times n_D \times n_F},$$

i.e. $$v_r = \frac{\text{product of number of teeth in drivers}}{\text{product of number of teeth in followers}} \dots (12).$$

This formula can be used for a compound train of any number of pairs.

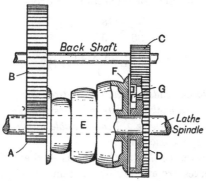

Fig. 182.   Back Gear Drive of a Lathe Spindle.

Sometimes, as in the back gear drive of a lathe spindle, Fig. 182, the driver and the follower are arranged on the same shaft, one of them being loosely mounted. In the normal working of the lathe the back shaft is moved backwards a little to bring the wheels out of gear and the cone pulley, which runs loosely upon the lathe spindle and has the wheel A integrally connected to it, is connected by a radially movable pin G, which enters between projections F on the pulley, the wheel D being keyed to the lathe spindle. To bring the back shaft into operation the pin G is released and the back gear is brought into engagement as shown. The same rule is used for the velocity ratio, no matter how the shafts are arranged.

Numerical Examples on Compound Trains.

(1) *In a compound gear train the driver A has* 40 *teeth and gears with a wheel B with* 20 *teeth. Keyed on the same shaft as B is a wheel C of* 120 *teeth gearing with the follower D with* 75 *teeth. If the wheel A runs at* 35 *revolutions per minute, how many revolutions per minute will the wheel D make ?*

In this case

$$v_r = \frac{\text{product of teeth in drivers}}{\text{product of teeth in followers}}$$

$$= \frac{40 \times 120}{20 \times 75}$$

$$= \frac{48}{15}.$$

$$\therefore \frac{\text{number of revolutions of follower}}{\text{number of revolutions of driver}} = \frac{48}{15},$$

$$\therefore \frac{N_D}{35} = \frac{48}{15},$$

$$\therefore N_D = \frac{48 \times 35}{15} = 112.$$

(2) *In a lathe headstock the lowest direct drive is* 50 *revolutions per minute and the back gear has to be designed so as to reduce this to* 5 *revolutions per minute. Find suitable numbers of teeth for the various wheels of the back gear.*

In this case we see that as the wheels A, B and C, D each form a pair whose axes are at the same distance apart, the sum of the radii of each pair of wheels must be the same, that is to say the sum of the number of teeth must be the same for each pair, if the teeth have the same pitch. In this case

$$v_R = \frac{5}{50} = \frac{1}{10},$$

$\therefore$ we have
$$\frac{1}{10} = \frac{n_A \times n_C}{n_B \times n_D}.$$

Further
$$n_A + n_B = n_C + n_D.$$

Suppose
$$n_A = 20.$$

Then if we take $n_B = 40$, $n_C = 10$, and $n_D = 50$, this gives us

$$v_R = \frac{20 \times 10}{40 \times 50} = \frac{1}{10}.$$

Reversing Tooth Drive for Lathe Lead Screw. The following arrangement of gearing is commonly employed for driving the lead screw L (Fig. 183) of a lathe from the headstock spindle O. The headstock spindle carries a toothed wheel A which drives a pinion D mounted on a spindle X either directly through a pinion B or else through a pinion C. The pinions B, C are mounted in a plate E pivoted on the shaft of the pinion D and provided with a slot engaging a stop-pin F for fixing it in its extreme positions. In the position (a) shown in the figure the pinion D is rotated in

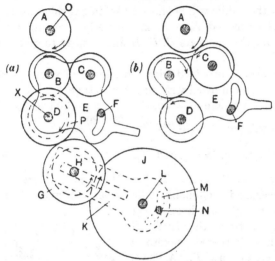

Fig. 183.   Reversing Drive for Lathe Lead Screw.

the same direction as the pinion A whereas in the position (b) the pinion D is driven in the opposite direction to the pinion A. In this position the pinion B has gone out of contact with the pinion A and the pinion C has come into contact with it. From the spindle X the drive goes through change wheels P, G, H and J adjustably carried in an arm K. The wheel H engages a wheel J on the lead screw shaft. The sizes of the wheels G, H and J are so chosen as to give the required velocity ratio. The spindle of the pinions G, H is adjusted in the slot in the quadrant so that the wheel H meshes correctly with the wheel J and the quadrant is then adjusted by means of a curved slot M so as to bring the wheel G into correct mesh with the wheel P. The quadrant is kept in its adjusted position by means of a locking bolt N.

The number of teeth on the wheel D is usually equal to that on the wheel A so that the spindle X rotates at the same speed as the headstock spindle.

Numerical Example. *The leading screw of a lathe is $\frac{3}{4}''$ pitch and it is required to cut a screw of 10 threads per inch. Find suitable sizes of the gear wheels.*

For one revolution of the lead screw the lathe saddle will be moved $\frac{3}{4}$ in., but for one revolution of the lathe spindle we wish the saddle to be moved only $\frac{1}{10}$ in.

Now the saddle will be moved 1 in. in $\frac{4}{3}$ revolutions and therefore will be moved $\frac{1}{10}$ in. in $\frac{4}{30}$ revolutions so that we have to choose our change gears P, G, H and J so as to give a velocity ratio of $\frac{4}{30} = \frac{2}{15}$.

The lathe is provided with a whole set of wheels of different numbers of teeth usually rising five at a time.

Suppose we take $n_P = 30$, $n_G = 75$, $n_H = 30$ and $n_J = 90$.

This will give $v_R = \dfrac{n_P \cdot n_H}{n_G \cdot n_J} = \dfrac{30 \times 30}{75 \times 90} = \dfrac{2}{15}$,

and this is the ratio required.

Bevel Gear Reversing Train. The following arrangement of bevel gearing is commonly adopted as a convenient reversing mechanism for a shaft. The drive goes from the shaft A which has feathered thereto a double clutch jaw F. A bevel wheel C is loosely mounted on the shaft A and engages a bevel wheel D, the other end of which engages a bevel wheel E fixed to a shaft B in line with the shaft A.

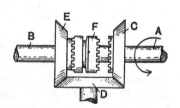

Fig. 184.   Bevel Gear Reversing Train.

The bevel wheels E and C each carry clutch jaws for engagement with the jaws on the clutch jaw F. In the position shown the rotation of the shaft A is transmitted direct from the clutch jaw to the bevel wheel E and thus to the shaft B, the wheels D and C rotating idly. The shafts A and B then rotate in the same direction. If the clutch F is moved to the right so as to engage with the wheel C the drive goes through the wheels C, D and E to the shaft B which then rotates in an opposite direction to the shaft A.

SUMMARY OF CHAPTER XVII.

Belt Gearing. H.P. transmitted $= \dfrac{\pi D_X N_X (T_1 - T_2)}{33,000}$.

T_1 may be taken as $2T_2$ in the absence of more exact information.

$$\text{Velocity ratio} = \frac{\text{diameter of driver}}{\text{diameter of follower}} \, .$$

In " open " belt the driver and follower rotate in the same direction, and in "crossed" belt they rotate in opposite directions.

Cone Pulleys. *Open belts.* If S is the distance apart of the shafts, the quantity $\dfrac{\pi}{2}(D_X + D_Y) + \dfrac{(D_X - D_Y)^2}{4S}$ must be constant.

Crossed belts. The sum of the diameters of corresponding pulleys must be constant.

Toothed Gearing.

$$\text{Diametral pitch} = \frac{\text{circular pitch}}{\pi} = \frac{\text{diameter of pitch circle}}{\text{number of teeth}} \, .$$

$$\text{Circular pitch} = \frac{\text{circumference of pitch circle}}{\text{number of teeth}} \, .$$

$$\text{Velocity ratio} = \frac{\text{number of teeth on driver}}{\text{number of teeth on follower}} \, .$$

Velocity ratio of compound train

$$= \frac{\text{product of number of teeth on drivers}}{\text{product of number of teeth on followers}} \, .$$

Idle wheels only alter the direction of rotation; they do not affect the velocity ratio.

EXERCISES. XVII.

1. A shaft is to be driven at 400 revolutions per min. and carries a pulley of 8 ins. diameter. What size driving pulley is necessary for a shaft which has to be driven from it at 70 revolutions per minute?

2. Two shafts at right angles to each other have to be driven by bevel gearing, the driving shaft runs at 120 revolutions per min. and carries a wheel with 48 teeth on it. How many teeth must be placed upon the second wheel if its shaft has to run at 320 revolutions per minute?

3. If a belt transmits 25 H.P. at 150 revolutions per minute over a pulley 3 ft. diameter find the difference of tension on the two sides of the belt.

If the tension on the tight side is three times that on the slack side find the tension on each side.

4. The crank of an engine is 2 ft. in length, and the diameter of the flywheel is 10 ft., also the flywheel has teeth on its rim and drives a pinion 3 ft. in diameter. If the mean pressure on the crank pin is $7\frac{1}{2}$ tons, what is the mean driving pressure on the teeth of the pinion?

5. A friction wheel 4 ft. diameter running at 70 revs. per min. drives a wheel 2 ft. 3 ins. diameter. Find the force with which the wheels must be pressed together per H.P. transmitted when the coefficient of friction for the surfaces is ·15.

6. In a lifting crab the length of the handle is 16 ins. and diameter of barrel 8 ins. The pinion on the same axis as the handle has 16 teeth, and gears with the spur wheel connected to the barrel which has 90 teeth. What weight can one man exerting a push of 30 lbs. lift?

7. The preceding is fitted to act with an increased velocity ratio by sliding the pinion out of contact with the spur wheel, and putting in gear a pinion of 18 teeth working with a spur wheel of 54 teeth. On the axis of the latter is another pinion of 18 teeth which now drives the 90 wheel. Find the force required to lift 1 ton.

8. The annexed sketch (Fig. XVII a) shows the arrangement of pulleys and belts used for driving a dynamo machine F from the steam-engine A.

$$\text{Diameter of } A = 57'', \quad B = 36'',$$
$$\text{,,} \quad \text{,,} \quad C = 42'', \quad D = 24'',$$
$$\text{,,} \quad \text{,,} \quad E = 48'', \quad F = 14''.$$

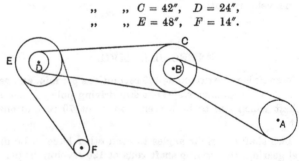

Fig. XVII a.

If the speed of A is 96 revs. per min. find the speed of F, assuming there is no slipping of belts.

9. A machine is driven from a pulley 4 ft. in diameter by means of a belt. If the difference of pull in the two sides of the belt is 20 lbs. weight, and the pulley makes 120 revolutions per min., find the H.P. transmitted by the belt.

10. The saddle of a lathe weighs 5 cwt.; it is moved along the bed by a rack and pinion arrangement. What force applied at the end of a handle 10″ long will be capable of just moving the saddle, supposing the pinion to have 12 teeth of $1\frac{1}{4}''$ pitch and the coefficient of friction between the saddle and the lathe bed to be ·1, other friction being neglected?

11. A leather belt $\frac{1}{4}$ inch thick has to transmit 10 H.P. from a pulley 4 ft. in diam. making 120 revolutions per minute. Assuming that the tension on the tight side is twice that on the slack side find the width of belt necessary if the safe stress in the belt is 320 lbs. per sq. in.

12. The tension per inch width of a belt must not exceed 110 lbs. Find the width required to transmit 12 H.P. from a shaft running at 80 revolutions per minute.

$$D = 4 \text{ ft. } 6 \text{ ins., } \frac{T_1}{T_2} = 1\frac{3}{4}.$$

13. A pulley 4 ft. in diameter is driven by two belts running over each other, each $\frac{3}{4}$ in. thick. The speed of the middle plane of the inner belt is 1800 ft. per minute. How much does the outer gain on the inner per minute?

14. The set of wheels for a screw cutting lathe range from 20 to 150 teeth, there being two 20 wheels. The leading screw has two threads to the inch. Arrange suitable trains for cutting threads on a $\frac{1}{4}$ in. screw, 20 threads to the inch.

15. The greatest and least diameters of the pulleys of a speed-cone for a headstock mandrel are 10″ and $5\frac{1}{2}''$ respectively; and this speed-cone is driven from a similar speed-cone keyed to a counter-shaft which makes 250 turns per min. The back gearing is of the usual type, the spur wheels concentric with the headstock spindle having 62 and 30 teeth gearing with wheels having 18 and 50 teeth respectively on the back spindle. Find the greatest and least revolutions per min. at which the headstock mandrel may be driven.

16. The effective diameter of a worm is 6″ and the pitch of the thread of the worm $2\frac{1}{2}''$. The worm is secured on the shaft of an engine of 60 B.H.P. and gears with a wheel on a shaft whose axis is at right angles to that of the engine shaft. If $\mu = \cdot16$ find η and H.P. transmitted by second shaft.

APPENDIX

THE SUM CURVE CONSTRUCTION

THE sum curve can be obtained graphically as follows. Let *ACD*, Fig. *a*, be any primitive curve on a straight base *AB*. Divide *AB* into any number of parts, not necessarily equal (but for convenience of working they are generally taken as equal).

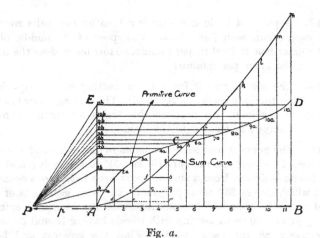

Fig. *a*.

These so-called base elements should be taken so small that the portion of the curve above them may be taken as a straight line. About 1 cm. or ·4 in. will usually be a suitable size and in most cases a smaller element, 11, will come at the end. Find the mid-points, 1, 2, 3, etc., of each of the base elements and let the verticals through these mid-points meet the curve in $1a$, $2a$, $3a$, etc. Now project the points on to a vertical line AE, thus obtaining the points $1b$, $2b$, $3b$, etc., and join such points to a pole P on AB produced and at some convenient distance p from A. Across space 1 then draw Ad parallel to $P1b$, de across space 2 parallel to $P2b$, and so on, until the point n is reached. Then the curve $Ade...n$ is the sum curve of the given curve, and to some scale Bn represents the area of the whole curve.

PROOF. Consider one of the elements, say 4, and draw fo horizontally.

Now $\triangle fgo$ is similar to the $\triangle P4bA$,

$$\therefore \frac{go}{fo} = \frac{4b, A}{PA},$$

but
$$PA = p \text{ and } 4b, A = 4, 4a,$$

$$\therefore go = \frac{fo \times 4, 4a}{p} = \frac{\text{area of element 4 of curve}}{p}.$$

Similarly $fq = \dfrac{\text{area of element 3 of curve}}{p}$ and so on,

$$\therefore \text{ ordinate through } g = go + fq + \ldots$$

$$= \frac{\text{area of first four elements of curve}}{p}.$$

$\therefore$ the curve $Ade...n$ is the sum curve required.

Then if Bn be measured on the vertical scale and p be measured on the horizontal scale, the area of the whole curve will be equal to $p \times Bn$.

It is obviously advisable to make p some convenient round number of units.

The sum curve obtained by this method may have the same operation performed on it, and thus the second sum curve of the primitive curve is obtained, and so on.

If the operation be performed on a rectangle, the sum curve will obviously become a sloping straight line, and if the sum curve of a sloping straight line be drawn, it will be found to be a parabola. In the case in which it is required to apply this construction to a curve which is not on a straight base, the curve is first brought to a straight base as follows:

Suppose $AcBd$, Fig. b, is a closed curve. Draw verticals through AB to meet a horizontal base $A'B'$. Divide the curve into a number of segments by vertical lines at short distances apart, and set up from the base $A'B'$ lengths a_1, b_1, etc., equal to

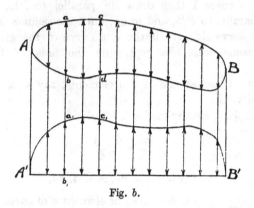

Fig. b.

the vertical portions a, b, etc., on the curve. Joining up the points thus obtained we get the corresponding curve $A'c_1B'$ on a straight base.

RIGHT-ANGLED TRIANGLES

$$\sin A = \frac{a}{b} \qquad \sec A = \frac{b}{c} \qquad \tan A = \frac{a}{c}$$

$$\cos A = \frac{c}{b} \qquad \operatorname{cosec} A = \frac{b}{a} \qquad \operatorname{cotan} A = \frac{c}{a}$$

Complement of $\theta = 90° - \theta$
Supplement of $\theta = 180° - \theta$

$$\tan \theta = \frac{\sin \theta}{\cos \theta}$$

$$\sin^2 \theta + \cos^2 \theta = 1$$

$$1 + \tan^2 \theta = \sec^2 \theta$$

versine $\theta = 1 - \cos \theta$

coversine $\theta = 1 - \sin \theta$

$$\sin (\theta + \phi) = \sin \theta \cos \phi + \cos \theta \sin \phi$$

$$\cos (\theta + \phi) = \cos \theta \cos \phi - \sin \theta \sin \phi$$

$$\tan (\theta + \phi) = \frac{\tan \theta + \tan \phi}{1 - \tan \theta \tan \phi}$$

$$\sin 2\theta = 2 \sin \theta \cos \theta$$

$$\cos 2\theta = \cos^2 \theta - \sin^2 \theta$$

$$\tan 2\theta = \frac{2 \tan \theta}{1 - \tan^2 \theta}$$

Given	Required	Formulae		
a, b	A, C, c	$\sin A = \dfrac{a}{b}$	$\cos C = \dfrac{a}{b}$	$c = \sqrt{(b+a)(b-a)}$
a, c	A, C, b	$\tan A = \dfrac{a}{c}$	$\operatorname{cotan} C = \dfrac{a}{c}$	$b = \sqrt{(a^2 + c^2)}$
A, a	C, c, b	$C = 90° - A$	$c = a \times \operatorname{cotan} A$	$b = \dfrac{a}{\sin A}$
A, b	C, a, c	$C = 90° - A$	$a = b \times \sin A$	$c = b \times \cos A$
A, c	C, a, b	$C = 90° - A$	$a = c \times \tan A$	$b = \dfrac{c}{\cos A}$

OBLIQUE-ANGLED TRIANGLES

$$s = \tfrac{1}{2}(a + b + c)$$

Given	Formulae
A, B, C, a A, b, c a, b, c	$\text{Area} = \begin{cases} (a^2 \times \sin B \times \sin C) \div 2 \sin A \\ \tfrac{1}{2}(c \times b \times \sin A) \\ \sqrt{s(s-a)(s-b)(s-c)} \end{cases}$

Given	Required	Formulae
A, C, a	c	$c = a \dfrac{\sin C}{\sin A}$
A, a, c	C	$\sin C = \dfrac{c \sin A}{a}$
a, c, B	A	$\tan A = \dfrac{a \sin B}{c - a \cos B}$
a, b, c	A	$\begin{cases} \sin \tfrac{1}{2}A = \sqrt{\dfrac{(s-b)(s-c)}{b \times c}} \\ \cos \tfrac{1}{2}A = \sqrt{\dfrac{s(s-a)}{b \times c}}; \quad \tan \tfrac{1}{2}A = \sqrt{\dfrac{(s-b)(s-c)}{s(s-a)}} \end{cases}$

LOGARITHMS.

	0	1	2	3	4	5	6	7	8	9	1	2	3	4	5	6	7	8	9
10	0000	0043	0086	0128	0170	0212	0253	0294	0334	0374	4	8	12	17	21	25	29	33	37
11	˙0414	0453	0492	0531	0569	0607	0645	0682	0719	0755	4	8	11	15	19	23	26	30	34
12	˙0792	0828	0864	0899	0934	0969	1004	1038	1072	1106	3	7	10	14	17	21	24	28	31
13	˙1139	1173	1206	1239	1271	1303	1335	1367	1399	1430	3	6	10	13	16	19	23	26	29
14	˙1461	1492	1523	1553	1584	1614	1644	1673	1703	1732	3	6	9	12	15	18	21	24	27
15	˙1761	1790	1818	1847	1875	1903	1931	1959	1987	2014	3	6	8	11	14	17	20	22	25
16	˙2041	2068	2095	2122	2148	2175	2201	2227	2253	2279	3	5	8	11	13	16	18	21	24
17	˙2304	2330	2355	2380	2405	2430	2455	2480	2504	2529	2	5	7	10	12	15	17	20	22
18	˙2553	2577	2601	2625	2648	2672	2695	2718	2742	2765	2	5	7	9	12	14	16	19	21
19	˙2788	2810	2833	2856	2878	2900	2923	2945	2967	2989	2	4	7	9	11	13	16	18	20
20	˙3010	3032	3054	3075	3096	3118	3139	3160	3181	3201	2	4	6	8	11	13	15	17	19
21	˙3222	3243	3263	3284	3304	3324	3345	3365	3385	3404	2	4	6	8	10	12	14	16	18
22	˙3424	3444	3464	3483	3502	3522	3541	3560	3579	3598	2	4	6	8	10	12	14	15	17
23	˙3617	3636	3655	3674	3692	3711	3729	3747	3766	3784	2	4	6	7	9	11	13	15	17
24	˙3802	3820	3838	3856	3874	3892	3909	3927	3945	3962	2	4	5	7	9	11	12	14	16
25	˙3979	3997	4014	4031	4048	4065	4082	4099	4116	4133	2	3	5	7	9	10	12	14	15
26	˙4150	4166	4183	4200	4216	4232	4249	4265	4281	4298	2	3	5	7	8	10	11	13	15
27	˙4314	4330	4346	4362	4378	4393	4409	4425	4440	4456	2	3	5	6	8	9	11	13	14
28	˙4472	4487	4502	4518	4533	4548	4564	4579	4594	4609	2	3	5	6	8	9	11	12	14
29	˙4624	4639	4654	4669	4683	4698	4713	4728	4742	4757	1	3	4	6	7	9	10	12	13
30	˙4771	4786	4800	4814	4829	4843	4857	4871	4886	4900	1	3	4	6	7	9	10	11	13
31	˙4914	4928	4942	4955	4969	4983	4997	5011	5024	5038	1	3	4	6	7	8	10	11	12
32	˙5051	5065	5079	5092	5105	5119	5132	5145	5159	5172	1	3	4	5	7	8	9	11	12
33	˙5185	5198	5211	5224	5237	5250	5263	5276	5289	5302	1	3	4	5	6	8	9	10	12
34	˙5315	5328	5340	5353	5366	5378	5391	5403	5416	5428	1	3	4	5	6	8	9	10	11
35	˙5441	5453	5465	5478	5490	5502	5514	5527	5539	5551	1	2	4	5	6	7	9	10	11
36	˙5563	5575	5587	5599	5611	5623	5635	5647	5658	5670	1	2	4	5	6	7	8	10	11
37	˙5682	5694	5705	5717	5729	5740	5752	5763	5775	5786	1	2	3	5	6	7	8	9	10
38	˙5798	5809	5821	5832	5843	5855	5866	5877	5888	5899	1	2	3	5	6	7	8	9	10
39	˙5911	5922	5933	5944	5955	5966	5977	5988	5999	6010	1	2	3	4	5	7	8	9	10
40	˙6021	6031	6042	6053	6064	6075	6085	6096	6107	6117	1	2	3	4	5	6	8	9	10
41	˙6128	6138	6149	6160	6170	6180	6191	6201	6212	6222	1	2	3	4	5	6	7	8	9
42	˙6232	6243	6253	6263	6274	6284	6294	6304	6314	6325	1	2	3	4	5	6	7	8	9
43	˙6335	6345	6355	6365	6375	6385	6395	6405	6415	6425	1	2	3	4	5	6	7	8	9
44	˙6435	6444	6454	6464	6474	6484	6493	6503	6513	6522	1	2	3	4	5	6	7	8	9
45	˙6532	6542	6551	6561	6571	6580	6590	6599	6609	6618	1	2	3	4	5	6	7	8	8
46	˙6628	6637	6646	6656	6665	6675	6684	6693	6702	6712	1	2	3	4	5	6	7	7	8
47	˙6721	6730	6739	6749	6758	6767	6776	6785	6794	6803	1	2	3	4	5	5	6	7	8
48	˙6812	6821	6830	6839	6848	6857	6866	6875	6884	6893	1	2	3	4	4	5	6	7	8
49	˙6902	6911	6920	6928	6937	6946	6955	6964	6972	6981	1	2	3	4	4	5	6	7	8
50	˙6990	6998	7007	7016	7024	7033	7042	7050	7059	7067	1	2	3	3	4	5	6	7	8
51	˙7076	7084	7093	7101	7110	7118	7126	7135	7143	7152	1	2	3	3	4	5	6	7	8
52	˙7160	7168	7177	7185	7193	7202	7210	7218	7226	7235	1	2	2	3	4	5	6	7	7
53	˙7243	7251	7259	7267	7275	7284	7292	7300	7308	7316	1	2	2	3	4	5	6	6	7
54	˙7324	7332	7340	7348	7356	7364	7372	7380	7388	7396	1	2	2	3	4	5	6	6	7

	0	1	2	3	4	5	6	7	8	9	1	2	3	4	5	6	7	8	9
55	˙7404	7412	7419	7427	7435	7443	7451	7459	7466	7474	1	2	2	3	4	5	5	6	7
56	˙7482	7490	7497	7505	7513	7520	7528	7536	7543	7551	1	2	2	3	4	5	5	6	7
57	˙7559	7566	7574	7582	7589	7597	7604	7612	7619	7627	1	2	2	3	4	5	5	6	7
58	˙7634	7642	7649	7657	7664	7672	7679	7686	7694	7701	1	1	2	3	4	4	5	6	7
59	˙7709	7716	7723	7731	7738	7745	7752	7760	7767	7774	1	1	2	3	4	4	5	6	7
60	˙7782	7789	7796	7803	7810	7818	7825	7832	7839	7846	1	1	2	3	4	4	5	6	6
61	˙7853	7860	7868	7875	7882	7889	7896	7903	7910	7917	1	1	2	3	4	4	5	6	6
62	˙7924	7931	7938	7945	7952	7959	7966	7973	7980	7987	1	1	2	3	3	4	5	6	6
63	˙7993	8000	8007	8014	8021	8028	8035	8041	8048	8055	1	1	2	3	3	4	5	5	6
64	˙8062	8069	8075	8082	8089	8096	8102	8109	8116	8122	1	1	2	3	3	4	5	5	6
65	˙8129	8136	8142	8149	8156	8162	8169	8176	8182	8189	1	1	2	3	3	4	5	5	6
66	˙8195	8202	8209	8215	8222	8228	8235	8241	8248	8254	1	1	2	3	3	4	5	5	6
67	˙8261	8267	8274	8280	8287	8293	8299	8306	8312	8319	1	1	2	3	3	4	5	5	6
68	˙8325	8331	8338	8344	8351	8357	8363	8370	8376	8382	1	1	2	3	3	4	4	5	6
69	˙8388	8395	8401	8407	8414	8420	8426	8432	8439	8445	1	1	2	2	3	4	4	5	6
70	˙8451	8457	8463	8470	8476	8482	8488	8494	8500	8506	1	1	2	2	3	4	4	5	6
71	˙8513	8519	8525	8531	8537	8543	8549	8555	8561	8567	1	1	2	2	3	4	4	5	5
72	˙8573	8579	8585	8591	8597	8603	8609	8615	8621	8627	1	1	2	2	3	4	4	5	5
73	˙8633	8639	8645	8651	8657	8663	8669	8675	8681	8686	1	1	2	2	3	4	4	5	5
74	˙8692	8698	8704	8710	8716	8722	8727	8733	8739	8745	1	1	2	2	3	4	4	5	5
75	˙8751	8756	8762	8768	8774	8779	8785	8791	8797	8802	1	1	2	2	3	3	4	5	5
76	˙8808	8814	8820	8825	8831	8837	8842	8848	8854	8859	1	1	2	2	3	3	4	5	5
77	˙8865	8871	8876	8882	8887	8893	8899	8904	8910	8915	1	1	2	2	3	3	4	4	5
78	˙8921	8927	8932	8938	8943	8949	8954	8960	8965	8971	1	1	2	2	3	3	4	4	5
79	˙8976	8982	8987	8993	8998	9004	9009	9015	9020	9025	1	1	2	2	3	3	4	4	5
80	˙9031	9036	9042	9047	9053	9058	9063	9069	9074	9079	1	1	2	2	3	3	4	4	5
81	˙9085	9090	9096	9101	9106	9112	9117	9122	9128	9133	1	1	2	2	3	3	4	4	5
82	˙9138	9143	9149	9154	9159	9165	9170	9175	9180	9186	1	1	2	2	3	3	4	4	5
83	˙9191	9196	9201	9206	9212	9217	9222	9227	9232	9238	1	1	2	2	3	3	4	4	5
84	˙9243	9248	9253	9258	9263	9269	9274	9279	9284	9289	1	1	2	2	3	3	4	4	5
85	˙9294	9299	9304	9309	9315	9320	9325	9330	9335	9340	1	1	2	2	3	3	4	4	5
86	˙9345	9350	9355	9360	9365	9370	9375	9380	9385	9390	1	1	1	2	3	3	4	4	5
87	˙9395	9400	9405	9410	9415	9420	9425	9430	9435	9440	0	1	1	2	2	3	3	4	4
88	˙9445	9450	9455	9460	9465	9469	9474	9479	9484	9489	0	1	1	2	2	3	3	4	4
89	˙9494	9499	9504	9509	9513	9518	9523	9528	9533	9538	0	1	1	2	2	3	3	4	4
90	˙9542	9547	9552	9557	9562	9566	9571	9576	9581	9586	0	1	1	2	2	3	3	4	4
91	˙9590	9595	9600	9605	9609	9614	9619	9624	9628	9633	0	1	1	2	2	3	3	4	4
92	˙9638	9643	9647	9652	9657	9661	9666	9671	9675	9680	0	1	1	2	2	3	3	4	4
93	˙9685	9689	9694	9699	9703	9708	9713	9717	9722	9727	0	1	1	2	2	3	3	4	4
94	˙9731	9736	9741	9745	9750	9754	9759	9763	9768	9773	0	1	1	2	2	3	3	4	4
95	˙9777	9782	9786	9791	9795	9800	9805	9809	9814	9818	0	1	1	2	2	3	3	4	4
96	˙9823	9827	9832	9836	9841	9845	9850	9854	9859	9863	0	1	1	2	2	3	3	4	4
97	˙9868	9872	9877	9881	9886	9890	9894	9899	9903	9908	0	1	1	2	2	3	3	4	4
98	˙9912	9917	9921	9926	9930	9934	9939	9943	9948	9952	0	1	1	2	2	3	3	4	4
99	˙9956	9961	9965	9969	9974	9978	9983	9987	9991	9996	0	1	1	2	2	3	3	3	4

300

ANTI-LOGARITHMS.

	0	1	2	3	4	5	6	7	8	9	1	2	3	4	5	6	7	8	9
·00	1000	1002	1005	1007	1009	1012	1014	1016	1019	1021	0	0	1	1	1	1	2	2	2
·01	1023	1026	1028	1030	1033	1035	1038	1040	1042	1045	0	0	1	1	1	1	2	2	2
·02	1047	1050	1052	1054	1057	1059	1062	1064	1067	1069	0	0	1	1	1	1	2	2	2
·03	1072	1074	1076	1079	1081	1084	1086	1089	1091	1094	0	0	1	1	1	1	2	2	2
·04	1096	1099	1102	1104	1107	1109	1112	1114	1117	1119	0	1	1	1	1	2	2	2	2
·05	1122	1125	1127	1130	1132	1135	1138	1140	1143	1146	0	1	1	1	1	2	2	2	2
·06	1148	1151	1153	1156	1159	1161	1164	1167	1169	1172	0	1	1	1	1	2	2	2	2
·07	1175	1178	1180	1183	1186	1189	1191	1194	1197	1199	0	1	1	1	1	2	2	2	2
·08	1202	1205	1208	1211	1213	1216	1219	1222	1225	1227	0	1	1	1	1	2	2	2	3
·09	1230	1233	1236	1239	1242	1245	1247	1250	1253	1256	0	1	1	1	1	2	2	2	3
·10	1259	1262	1265	1268	1271	1274	1276	1279	1282	1285	0	1	1	1	1	2	2	2	3
·11	1288	1291	1294	1297	1300	1303	1306	1309	1312	1315	0	1	1	1	2	2	2	2	3
·12	1318	1321	1324	1327	1330	1334	1337	1340	1343	1346	0	1	1	1	2	2	2	2	3
·13	1349	1352	1355	1358	1361	1365	1368	1371	1374	1377	0	1	1	1	2	2	2	3	3
·14	1380	1384	1387	1390	1393	1396	1400	1403	1406	1409	0	1	1	1	2	2	2	3	3
·15	1413	1416	1419	1422	1426	1429	1432	1435	1439	1442	0	1	1	1	2	2	2	3	3
·16	1445	1449	1452	1455	1459	1462	1466	1469	1472	1476	0	1	1	1	2	2	2	3	3
·17	1479	1483	1486	1489	1493	1496	1500	1503	1507	1510	0	1	1	1	2	2	2	3	3
·18	1514	1517	1521	1524	1528	1531	1535	1538	1542	1545	0	1	1	1	2	2	2	3	3
·19	1549	1552	1556	1560	1563	1567	1570	1574	1578	1581	0	1	1	1	2	2	3	3	3
·20	1585	1589	1592	1596	1600	1603	1607	1611	1614	1618	0	1	1	1	2	2	3	3	3
·21	1622	1626	1629	1633	1637	1641	1644	1648	1652	1656	0	1	1	2	2	2	3	3	3
·22	1660	1663	1667	1671	1675	1679	1683	1687	1690	1694	0	1	1	2	2	2	3	3	3
·23	1698	1702	1706	1710	1714	1718	1722	1726	1730	1734	0	1	1	2	2	2	3	3	4
·24	1738	1742	1746	1750	1754	1758	1762	1766	1770	1774	0	1	1	2	2	2	3	3	4
·25	1778	1782	1786	1791	1795	1799	1803	1807	1811	1816	0	1	1	2	2	2	3	3	4
·26	1820	1824	1828	1832	1837	1841	1845	1849	1854	1858	0	1	1	2	2	3	3	3	4
·27	1862	1866	1871	1875	1879	1884	1888	1892	1897	1901	0	1	1	2	2	3	3	3	4
·28	1905	1910	1914	1919	1923	1928	1932	1936	1941	1945	0	1	1	2	2	3	3	4	4
·29	1950	1954	1959	1963	1968	1972	1977	1982	1986	1991	0	1	1	2	2	3	3	4	4
·30	1995	2000	2004	2009	2014	2018	2023	2028	2032	2037	0	1	1	2	2	3	3	4	4
·31	2042	2046	2051	2056	2061	2065	2070	2075	2080	2084	0	1	1	2	2	3	3	4	4
·32	2089	2094	2099	2104	2109	2113	2118	2123	2128	2133	0	1	1	2	2	3	3	4	4
·33	2138	2143	2148	2153	2158	2163	2168	2173	2178	2183	0	1	1	2	2	3	3	4	4
·34	2188	2193	2198	2203	2208	2213	2218	2223	2228	2234	1	1	2	2	3	3	4	4	5
·35	2239	2244	2249	2254	2259	2265	2270	2275	2280	2286	1	1	2	2	3	3	4	4	5
·36	2291	2296	2301	2307	2312	2317	2323	2328	2333	2339	1	1	2	2	3	3	4	4	5
·37	2344	2350	2355	2360	2366	2371	2377	2382	2388	2393	1	1	2	2	3	3	4	4	5
·38	2399	2404	2410	2415	2421	2427	2432	2438	2443	2449	1	1	2	2	3	3	4	4	5
·39	2455	2460	2466	2472	2477	2483	2489	2495	2500	2506	1	1	2	2	3	3	4	5	5
·40	2512	2518	2523	2529	2535	2541	2547	2553	2559	2564	1	1	2	2	3	4	4	5	5
·41	2570	2576	2582	2588	2594	2600	2606	2612	2618	2624	1	1	2	2	3	4	4	5	5
·42	2630	2636	2642	2649	2655	2661	2667	2673	2679	2685	1	1	2	2	3	4	4	5	6
·43	2692	2698	2704	2710	2716	2723	2729	2735	2742	2748	1	1	2	3	3	4	4	5	6
·44	2754	2761	2767	2773	2780	2786	2793	2799	2805	2812	1	1	2	3	3	4	4	5	6
·45	2818	2825	2831	2838	2844	2851	2858	2864	2871	2877	1	1	2	3	3	4	5	5	6
·46	2884	2891	2897	2904	2911	2917	2924	2931	2938	2944	1	1	2	3	3	4	5	5	6
·47	2951	2958	2965	2972	2979	2985	2992	2999	3006	3013	1	1	2	3	3	4	5	5	6
·48	3020	3027	3034	3041	3048	3055	3062	3069	3076	3083	1	1	2	3	4	4	5	6	6
·49	3090	3097	3105	3112	3119	3126	3133	3141	3148	3155	1	1	2	3	4	4	5	6	6

	0	1	2	3	4	5	6	7	8	9	1	2	3	4	5	6	7	8	9
50	3162	3170	3177	3184	3192	3199	3206	3214	3221	3228	1	1	2	3	4	4	5	6	7
·51	3236	3243	3251	3258	3266	3273	3281	3289	3296	3304	1	2	2	3	4	5	5	6	7
·52	3311	3319	3327	3334	3342	3350	3357	3365	3373	3381	1	2	2	3	4	5	5	6	7
·53	3388	3396	3404	3412	3420	3428	3436	3443	3451	3459	1	2	2	3	4	5	6	6	7
·54	3467	3475	3483	3491	3499	3508	3516	3524	3532	3540	1	2	2	3	4	5	6	6	7
·55	3548	3556	3565	3573	3581	3589	3597	3606	3614	3622	1	2	2	3	4	5	6	7	7
·56	3631	3639	3648	3656	3664	3673	3681	3690	3698	3707	1	2	3	3	4	5	6	7	8
·57	3715	3724	3733	3741	3750	3758	3767	3776	3784	3793	1	2	3	3	4	5	6	7	8
·58	3802	3811	3819	3828	3837	3846	3855	3864	3873	3882	1	2	3	4	4	5	6	7	8
·59	3890	3899	3908	3917	3926	3936	3945	3954	3963	3972	1	2	3	4	5	5	6	7	8
60	3981	3990	3999	4009	4018	4027	4036	4046	4055	4064	1	2	3	4	5	6	6	7	8
·61	4074	4083	4093	4102	4111	4121	4130	4140	4150	4159	1	2	3	4	5	6	7	8	9
·62	4169	4178	4188	4198	4207	4217	4227	4236	4246	4256	1	2	3	4	5	6	7	8	9
·63	4266	4276	4285	4295	4305	4315	4325	4335	4345	4355	1	2	3	4	5	6	7	8	9
·64	4365	4375	4385	4395	4406	4416	4426	4436	4446	4457	1	2	3	4	5	6	7	8	9
·65	4467	4477	4487	4498	4508	4519	4529	4539	4550	4560	1	2	3	4	5	6	7	8	9
·66	4571	4581	4592	4603	4613	4624	4634	4645	4656	4667	1	2	3	4	5	6	7	9	10
·67	4677	4688	4699	4710	4721	4732	4742	4753	4764	4775	1	2	3	4	5	7	8	9	10
·68	4786	4797	4808	4819	4831	4842	4853	4864	4875	4887	1	2	3	4	6	7	8	9	10
·69	4898	4909	4920	4932	4943	4955	4966	4977	4989	5000	1	2	3	5	6	7	8	9	10
·70	5012	5023	5035	5047	5058	5070	5082	5093	5105	5117	1	2	4	5	6	7	8	9	11
·71	5129	5140	5152	5164	5176	5188	5200	5212	5224	5236	1	2	4	5	6	7	8	10	11
·72	5248	5260	5272	5284	5297	5309	5321	5333	5346	5358	1	2	4	5	6	7	9	10	11
·73	5370	5383	5395	5408	5420	5433	5445	5458	5470	5483	1	3	4	5	6	8	9	10	11
·74	5495	5508	5521	5534	5546	5559	5572	5585	5598	5610	1	3	4	5	6	8	9	10	12
·75	5623	5636	5649	5662	5675	5689	5702	5715	5728	5741	1	3	4	5	7	8	9	10	12
·76	5754	5768	5781	5794	5808	5821	5834	5848	5861	5875	1	3	4	5	7	8	9	11	12
·77	5888	5902	5916	5929	5943	5957	5970	5984	5998	6012	1	3	4	5	7	8	10	11	12
·78	6026	6039	6053	6067	6081	6095	6109	6124	6138	6152	1	3	4	6	7	8	10	11	13
·79	6166	6180	6194	6209	6223	6237	6252	6266	6281	6295	1	3	4	6	7	9	10	11	13
·80	6310	6324	6339	6353	6368	6383	6397	6412	6427	6442	1	3	4	6	7	9	10	12	13
·81	6457	6471	6486	6501	6516	6531	6546	6561	6577	6592	2	3	5	6	8	9	11	12	14
·82	6607	6622	6637	6653	6668	6683	6699	6714	6730	6745	2	3	5	6	8	9	11	12	14
·83	6761	6776	6792	6808	6823	6839	6855	6871	6887	6902	2	3	5	6	8	9	11	13	14
·84	6918	6934	6950	6966	6982	6998	7015	7031	7047	7063	2	3	5	6	8	10	11	13	15
·85	7079	7096	7112	7129	7145	7161	7178	7194	7211	7228	2	3	5	7	8	10	12	13	15
·86	7244	7261	7278	7295	7311	7328	7345	7362	7379	7396	2	3	5	7	8	10	12	13	15
·87	7413	7430	7447	7464	7482	7499	7516	7534	7551	7568	2	3	5	7	9	10	12	14	16
·88	7586	7603	7621	7638	7656	7674	7691	7709	7727	7745	2	4	5	7	9	11	12	14	16
·89	7762	7780	7798	7816	7834	7852	7870	7889	7907	7925	2	4	5	7	9	11	13.14	16
·90	7943	7962	7980	7998	8017	8035	8054	8072	8091	8110	2	4	6	7	9	11	13	15	17
·91	8128	8147	8166	8185	8204	8222	8241	8260	8279	8299	2	4	6	8	9	11	13	15	17
·92	8318	8337	8356	8375	8395	8414	8433	8453	8472	8492	2	4	6	8	10	12	14	15	17
·93	8511	8531	8551	8570	8590	8610	8630	8650	8670	8690	2	4	6	8	10	12	14	16	18
·94	8710	8730	8750	8770	8790	8810	8831	8851	8872	8892	2	4	6	8	10	12	14	16	18
·95	8913	8933	8954	8974	8995	9016	9036	9057	9078	9099	2	4	6	8	10	12	15	17	19
·96	9120	9141	9162	9183	9204	9226	9247	9268	9290	9311	2	4	6	8	11	13	15	17	19
·97	9333	9354	9376	9397	9419	9441	9462	9484	9506	9528	2	4	7	9	11	13	15	17	20
·98	9550	9572	9594	9616	9638	9661	9683	9705	9727	9750	2	4	7	9	11	13	16	18	20
·99	9772	9795	9817	9840	9863	9886	9908	9931	9954	9977	2	5	7	9	11	14	16	18	20

	O'	6'	12'	18'	24'	30'	36'	42'	48'	54'	1'	2'	3'	4'	5'
0	·0000	0017	0035	0052	0070	0087	0105	0122	0140	0157	3	6	9	12	15
1	·0175	0192	0209	0227	0244	0262	0279	0297	0314	0332	3	6	9	12	15
2	·0349	0366	0384	0401	0419	0436	0454	0471	0488	0506	3	6	9	12	15
3	·0523	0541	0558	0576	0593	0610	0628	0645	0663	0680	3	6	9	12	15
4	·0698	0715	0732	0750	0767	0785	0802	0819	0837	0854	3	6	9	12	14
5	·0872	0889	0906	0924	0941	0958	0976	0993	1011	1028	3	6	9	12	14
6	·1045	1063	1080	1097	1115	1132	1149	1167	1184	1201	3	6	9	12	14
7	·1219	1236	1253	1271	1288	1305	1323	1340	1357	1374	3	6	9	12	14
8	·1392	1409	1426	1444	1461	1478	1495	1513	1530	1547	3	6	9	12	14
9	·1564	1582	1599	1616	1633	1650	1668	1685	1702	1719	3	6	9	11	14
10	·1736	1754	1771	1788	1805	1822	1840	1857	1874	1891	3	6	9	11	14
11	·1908	1925	1942	1959	1977	1994	2011	2028	2045	2062	3	6	9	11	14
12	·2079	2096	2113	2130	2147	2164	2181	2198	2215	2233	3	6	9	11	14
13	·2250	2267	2284	2300	2317	2334	2351	2368	2385	2402	3	6	8	11	14
14	·2419	2436	2453	2470	2487	2504	2521	2538	2554	2571	3	6	8	11	14
15	·2588	2605	2622	2639	2656	2672	2689	2706	2723	2740	3	6	8	11	14
16	·2756	2773	2790	2807	2823	2840	2857	2874	2890	2907	3	6	8	11	14
17	·2924	2940	2957	2974	2990	3007	3024	3040	3057	3074	3	6	8	11	14
18	·3090	3107	3123	3140	3156	3173	3190	3206	3223	3239	3	6	8	11	14
19	·3256	3272	3289	3305	3322	3338	3355	3371	3387	3404	3	5	8	11	14
20	·3420	3437	3453	3469	3486	3502	3518	3535	3551	3567	3	5	8	11	14
21	·3584	3600	3616	3633	3649	3665	3681	3697	3714	3730	3	5	8	11	14
22	·3746	3762	3778	3795	3811	3827	3843	3859	3875	3891	3	5	8	11	13
23	·3907	3923	3939	3955	3971	3987	4003	4019	4035	4051	3	5	8	11	13
24	·4067	4083	4099	4115	4131	4147	4163	4179	4195	4210	3	5	8	11	13
25	·4226	4242	4258	4274	4289	4305	4321	4337	4352	4368	3	5	8	11	13
26	·4384	4399	4415	4431	4446	4462	4478	4493	4509	4524	3	5	8	10	13
27	·4540	4555	4571	4586	4602	4617	4633	4648	4664	4679	3	5	8	10	13
28	·4695	4710	4726	4741	4756	4772	4787	4802	4818	4833	3	5	8	10	13
29	·4848	4863	4879	4894	4909	4924	4939	4955	4970	4985	3	5	8	10	13
30	·5000	5015	5030	5045	5060	5075	5090	5105	5120	5135	3	5	8	10	13
31	·5150	5165	5180	5195	5210	5225	5240	5255	5270	5284	2	5	7	10	12
32	·5299	5314	5329	5344	5358	5373	5388	5402	5417	5432	2	5	7	10	12
33	·5446	5461	5476	5490	5505	5519	5534	5548	5563	5577	2	5	7	10	12
34	·5592	5606	5621	5635	5650	5664	5678	5693	5707	5721	2	5	7	10	12
35	·5736	5750	5764	5779	5793	5807	5821	5835	5850	5864	2	5	7	9	12
36	·5878	5892	5906	5920	5934	5948	5962	5976	5990	6004	2	5	7	9	12
37	·6018	6032	6046	6060	6074	6088	6101	6115	6129	6143	2	5	7	9	12
38	·6157	6170	6184	6198	6211	6225	6239	6252	6266	6280	2	5	7	9	11
39	·6293	6307	6320	6334	6347	6361	6374	6388	6401	6414	2	4	7	9	11
40	·6428	6441	6455	6468	6481	6494	6508	6521	6534	6547	2	4	7	9	11
41	·6561	6574	6587	6600	6613	6626	6639	6652	6665	6678	2	4	7	9	11
42	·6691	6704	6717	6730	6743	6756	6769	6782	6794	6807	2	4	6	9	11
43	·6820	6833	6845	6858	6871	6884	6896	6909	6921	6934	2	4	6	8	11
44	·6947	6959	6972	6984	6997	7009	7022	7034	7046	7059	2	4	6	8	10

	O′	6′	12′	18′	24′	30′	36′	42′	48′	54′	1′	2′	3′	4′	5′
45°	·7071	7083	7096	7108	7120	7133	7145	7157	7169	7181	2	4	6	8	10
46	·7193	7206	7218	7230	7242	7254	7266	7278	7290	7302	2	4	6	8	10
47	·7314	7325	7337	7349	7361	7373	7385	7396	7408	7420	2	4	6	8	10
48	·7431	7443	7455	7466	7478	7490	7501	7513	7524	7536	2	4	6	8	10
49	·7547	7559	7570	7581	7593	7604	7615	7627	7638	7649	2	4	6	8	9
50	·7660	7672	7683	7694	7705	7716	7727	7738	7749	7760	2	4	6	7	9
51	·7771	7782	7793	7804	7815	7826	7837	7848	7859	7869	2	4	5	7	9
52	·7880	7891	7902	7912	7923	7934	7944	7955	7965	7976	2	4	5	7	9
53	·7986	7997	8007	8018	8028	8039	8049	8059	8070	8080	2	3	5	7	9
54	·8090	8100	8111	8121	8131	8141	8151	8161	8171	8181	2	3	5	7	8
55	·8192	8202	8211	8221	8231	8241	8251	8261	8271	8281	2	3	5	7	8
56	·8290	8300	8310	8320	8329	8339	8348	8358	8368	8377	2	3	5	6	8
57	·8387	8396	8406	8415	8425	8434	8443	8453	8462	8471	2	3	5	6	8
58	·8480	8490	8499	8508	8517	8526	8536	8545	8554	8563	2	3	5	6	8
59	·8572	8581	8590	8599	8607	8616	8625	8634	8643	8652	1	3	4	6	7
60	·8660	8669	8678	8686	8695	8704	8712	8721	8729	8738	1	3	4	6	7
61	·8746	8755	8763	8771	8780	8788	8796	8805	8813	8821	1	3	4	6	7
62	·8829	8838	8846	8854	8862	8870	8878	8886	8894	8902	1	3	4	5	7
63	·8910	8918	8926	8934	8942	8949	8957	8965	8973	8980	1	3	4	5	6
64	·8988	8996	9003	9011	9018	9026	9033	9041	9048	9056	1	3	4	5	6
65	·9063	9070	9078	9085	9092	9100	9107	9114	9121	9128	1	2	4	5	6
66	·9135	9143	9150	9157	9164	9171	9178	9184	9191	9198	1	2	3.	5	6
67	·9205	9212	9219	9225	9232	9239	9245	9252	9259	9265	1	2	3	4	6
68	·9272	9278	9285	9291	9298	9304	9311	9317	9323	9330	1	2	3	4	5
69	·9336	9342	9348	9354	9361	9367	9373	9379	9385	9391	1	2	3	4	5
70	·9397	9403	9409	9415	9421	9426	9432	9438	9444	9449	1	2	3	4	5
71	·9455	9461	9466	9472	9478	9483	9489	9494	9500	9505	1	2	3	4	5
72	·9511	9516	9521	9527	9532	9537	9542	9548	9553	9558	1	2	3	4	4
73	·9563	9568	9573	9578	9583	9588	9593	9598	9603	9608	1	2	2	3	4
74	·9613	9617	9622	9627	9632	9636	9641	9646	9650	9655	1	2	2	3	4
75	·9659	9664	9668	9673	9677	9681	9686	9690	9694	9699	1	1	2	3	4
76	·9703	9707	9711	9715	9720	9724	9728	9732	9736	9740	1	1	2	3	3
77	·9744	9748	9751	9755	9759	9763	9767	9770	9774	9778	1	1	2	3	3
78	·9781	9785	9789	9792	9796	9799	9803	9806	9810	9813	1	1	2	2	3
79	·9816	9820	9823	9826	9829	9833	9836	9839	9842	9845	1	1	2	2	3
80	·9848	9851	9854	9857	9860	9863	9866	9869	9871	9874	0	1	1	2	2
81	·9877	9880	9882	9885	9888	9890	9893	9895	9898	9900	0	1	1	2	2
82	·9903	9905	9907	9910	9912	9914	9917	9919	9921	9923	0	1	1	2	2
83	·9925	9928	9930	9932	9934	9936	9938	9940	9942	9943	0	1	1	1	2
84	·9945	9947	9949	9951	9952	9954	9956	9957	9959	9960	0	1	1	1	1
85	·9962	9963	9965	9966	9968	9969	9971	9972	9973	9974	0	0	1	1	1
86	·9976	9977	9978	9979	9980	9981	9982	9983	9984	9985	0	0	1	1	1
87	·9986	9987	9988	9989	9990	9990	9991	9992	9993	9993					
88	·9994	9995	9995	9996	9996	9997	9997	9997	9998	9998					
89	·9998	9999	9999	9999	9999	1·000	1·000	1·000	1·000	1·000					

°	0′	6′	12′	18′	24′	30′	36′	42′	48′	54′	1′	2′	3′	4′	5′
0	1·0000	1·000	1·000	1·000	1·000	1·000	9999	9999	9999	9999					
1	·9998	9998	9998	9997	9997	9997	9996	9996	9995	9995					
2	·9994	9993	9993	9992	9991	9990	9990	9989	9988	9987					
3	·9986	9985	9984	9983	9982	9981	9980	9979	9978	9977	0	0	1	1	1
4	·9976	9974	9973	9972	9971	9969	9968	9966	9965	9963	0	0	1	1	1
5	·9962	9960	9959	9957	9956	9954	9952	9951	9949	9947	0	1	1	1	1
6	·9945	9943	9942	9940	9938	9936	9934	9932	9930	9928	0	1	1	1	2
7	·9925	9923	9921	9919	9917	9914	9912	9910	9907	9905	0	1	1	2	2
8	·9903	9900	9898	9895	9893	9890	9888	9885	9882	9880	0	1	1	2	2
9	·9877	9874	9871	9869	9866	9863	9860	9857	9854	9851	0	1	1	2	2
10	·9848	9845	9842	9839	9836	9833	9829	9826	9823	9820	1	1	2	2	3
11	·9816	9813	9810	9806	9803	9799	9796	9792	9789	9785	1	1	2	2	3
12	·9781	9778	9774	9770	9767	9763	9759	9755	9751	9748	1	1	2	3	3
13	·9744	9740	9736	9732	9728	9724	9720	9715	9711	9707	1	1	2	3	3
14	·9703	9699	9694	9690	9686	9681	9677	9673	9668	9664	1	1	2	3	4
15	·9659	9655	9650	9646	9641	9636	9632	9627	9622	9617	1	2	2	3	4
16	·9613	9608	9603	9598	9593	9588	9583	9578	9573	9568	1	2	2	3	4
17	·9563	9558	9553	9548	9542	9537	9532	9527	9521	9516	1	2	3	4	4
18	·9511	9505	9500	9494	9489	9483	9478	9472	9466	9461	1	2	3	4	5
19	·9455	9449	9444	9438	9432	9426	9421	9415	9409	9403	1	2	3	4	5
20	·9397	9391	9385	9379	9373	9367	9361	9354	9348	9342	1	2	3	4	5
21	·9336	9330	9323	9317	9311	9304	9298	9291	9285	9278	1	2	3	4	5
22	·9272	9265	9259	9252	9245	9239	9232	9225	9219	9212	1	2	3	4	6
23	·9205	9198	9191	9184	9178	9171	9164	9157	9150	9143	1	2	3	5	6
24	·9135	9128	9121	9114	9107	9100	9092	9085	9078	9070	1	2	4	5	6
25	·9063	9056	9048	9041	9033	9026	9018	9011	9003	8996	1	3	4	5	6
26	·8988	8980	8973	8965	8957	8949	8942	8934	8926	8918	1	3	4	5	6
27	·8910	8902	8894	8886	8878	8870	8862	8854	8846	8838	1	3	4	5	7
28	·8829	8821	8813	8805	8796	8788	8780	8771	8763	8755	1	3	4	6	7
29	·8746	8738	8729	8721	8712	8704	8695	8686	8678	8669	1	3	4	6	7
30	·8660	8652	8643	8634	8625	8616	8607	8599	8590	8581	1	3	4	6	7
31	·8572	8563	8554	8545	8536	8526	8517	8508	8499	8490	2	3	5	6	8
32	·8480	8471	8462	8453	8443	8434	8425	8415	8406	8396	2	3	5	6	8
33	·8387	8377	8368	8358	8348	8339	8329	8320	8310	8300	2	3	5	6	8
34	·8290	8281	8271	8261	8251	8241	8231	8221	8211	8202	2	3	5	7	8
35	·8192	8181	8171	8161	8151	8141	8131	8121	8111	8100	2	3	5	7	8
36	·8090	8080	8070	8059	8049	8039	8028	8018	8007	7997	2	3	5	7	9
37	·7986	7976	7965	7955	7944	7934	7923	7912	7902	7891	2	4	5	7	9
38	·7880	7869	7859	7848	7837	7826	7815	7804	7793	7782	2	4	5	7	9
39	·7771	7760	7749	7738	7727	7716	7705	7694	7683	7672	2	4	6	7	9
40	·7660	7649	7638	7627	7615	7604	7593	7581	7570	7559	2	4	6	8	9
41	·7547	7536	7524	7513	7501	7490	7478	7466	7455	7443	2	4	6	8	10
42	·7431	7420	7408	7396	7385	7373	7361	7349	7337	7325	2	4	6	8	10
43	·7314	7302	7290	7278	7266	7254	7242	7230	7218	7206	2	4	6	8	10
44	·7193	7181	7169	7157	7145	7133	7120	7108	7096	7083	2	4	6	8	10

The black type indicates that the integer changes.

	0'	6'	12'	18'	24'	30'	36'	42'	48'	54'	1'	2'	3'	4'	5'
45	·7071	7059	7046	7034	7022	7009	6997	6984	6972	6959	2	4	6	8	10
46	·6947	6934	6921	6909	6896	6884	6871	6858	6845	6833	2	4	6	8	11
47	·6820	6807	6794	6782	6769	6756	6743	6730	6717	6704	2	4	6	9	11
48	·6691	6678	6665	6652	6639	6626	6613	6600	6587	6574	2	4	7	9	11
49	·6561	6547	6534	6521	6508	6494	6481	6468	6455	6441	2	4	7	9	11
50	·6428	6414	6401	6388	6374	6361	6347	6334	6320	6307	2	4	7	9	11
51	·6293	6280	6266	6252	6239	6225	6211	6198	6184	6170	2	5	7	9	11
52	·6157	6143	6129	6115	6101	6088	6074	6060	6046	6032	2	5	7	9	12
53	·6018	6004	5990	5976	5962	5948	5934	5920	5906	5892	2	5	7	9	12
54	·5878	5864	5850	5835	5821	5807	5793	5779	5764	5750	2	5	7	9	12
55	·5736	5721	5707	5693	5678	5664	5650	5635	5621	5606	2	5	7	10	12
56	·5592	5577	5563	5548	5534	5519	5505	5490	5476	5461	2	5	7	10	12
57	·5446	5432	5417	5402	5388	5373	5358	5344	5329	5314	2	5	7	10	12
58	·5299	5284	5270	5255	5240	5225	5210	5195	5180	5165	2	5	7	10	12
59	·5150	5135	5120	5105	5090	5075	5060	5045	5030	5015	3	5	8	10	13
60	·5000	4985	4970	4955	4939	4924	4909	4894	4879	4863	3	5	8	10	13
61	·4848	4833	4818	4802	4787	4772	4756	4741	4726	4710	3	5	8	10	13
62	·4695	4679	4664	4648	4633	4617	4602	4586	4571	4555	3	5	8	10	13
63	·4540	4524	4509	4493	4478	4462	4446	4431	4415	4399	3	5	8	11	13
64	·4384	4368	4352	4337	4321	4305	4289	4274	4258	4242	3	5	8	11	13
65	·4226	4210	4195	4179	4163	4147	4131	4115	4099	4083	3	5	8	11	13
66	·4067	4051	4035	4019	4003	3987	3971	3955	3939	3923	3	5	8	11	13
67	·3907	3891	3875	3859	3843	3827	3811	3795	3778	3762	3	5	8	11	13
68	·3746	3730	3714	3697	3681	3665	3649	3633	3616	3600	3	5	8	11	14
69	·3584	3567	3551	3535	3518	3502	3486	3469	3453	3437	3	5	8	11	14
70	·3420	3404	3387	3371	3355	3338	3322	3305	3289	3272	3	5	8	11	14
71	·3256	3239	3223	3206	3190	3173	3156	3140	3123	3107	3	6	8	11	14
72	·3090	3074	3057	3040	3024	3007	2990	2974	2957	2940	3	6	8	11	14
73	·2924	2907	2890	2874	2857	2840	2823	2807	2790	2773	3	6	8	11	14
74	·2756	2740	2723	2706	2689	2672	2656	2639	2622	2605	3	6	8	11	14
75	·2588	2571	2554	2538	2521	2504	2487	2470	2453	2436	3	6	8	11	14
76	·2419	2402	2385	2368	2351	2334	2317	2300	2284	2267	3	6	8	11	14
77	·2250	2233	2215	2198	2181	2164	2147	2130	2113	2096	3	6	9	11	14
78	·2079	2062	2045	2028	2011	1994	1977	1959	1942	1925	3	6	9	11	14
79	·1908	1891	1874	1857	1840	1822	1805	1788	1771	1754	3	6	9	11	14
80	·1736	1719	1702	1685	1668	1650	1633	1616	1599	1582	3	6	9	11	14
81	·1564	1547	1530	1513	1495	1478	1461	1444	1426	1409	3	6	9	12	14
82	·1392	1374	1357	1340	1323	1305	1288	1271	1253	1236	3	6	9	12	14
83	·1219	1201	1184	1167	1149	1132	1115	1097	1080	1063	3	6	9	12	14
84	·1045	1028	1011	0993	0976	0958	0941	0924	0906	0889	3	6	9	12	14
85	·0872	0854	0837	0819	0802	0785	0767	0750	0732	0715	3	6	9	12	14
86	·0698	0680	0663	0645	0628	0610	0593	0576	0558	0541	3	6	9	12	15
87	·0523	0506	0488	0471	0454	0436	0419	0401	0384	0366	3	6	9	12	15
88	·0349	0332	0314	0297	0279	0262	0244	0227	0209	0192	3	6	9	12	15
89	·0175	0157	0140	0122	0105	0087	0070	0052	0035	0017	3	6	9	12	15

A. M.

	O'	6'	12'	18'	24'	30'	36'	42'	48'	54'	1'	2'	3'	4'	5'
0	0·0000	0017	0035	0052	0070	0087	0105	0122	0140	0157	3	6	9	12	15
1	0·0175	0192	0209	0227	0244	0262	0279	0297	0314	0332	3	6	9	12	15
2	0·0349	0367	0384	0402	0419	0437	0454	0472	0489	0507	3	6	9	12	15
3	0·0524	0542	0559	0577	0594	0612	0629	0647	0664	0682	3	6	9	12	15
4	0·0699	0717	0734	0752	0769	0787	0805	0822	0840	0857	3	6	9	12	15
5	0·0875	0892	0910	0928	0945	0963	0981	0998	1016	1033	3	6	9	12	15
6	0·1051	1069	1086	1104	1122	1139	1157	1175	1192	1210	3	6	9	12	15
7	0·1228	1246	1263	1281	1299	1317	1334	1352	1370	1388	3	6	9	12	15
8	0·1405	1423	1441	1459	1477	1495	1512	1530	1548	1566	3	6	9	12	15
9	0·1584	1602	1620	1638	1655	1673	1691	1709	1727	1745	3	6	9	12	15
10	0·1763	1781	1799	1817	1835	1853	1871	1890	1908	1926	3	6	9	12	15
11	0·1944	1962	1980	1998	2016	2035	2053	2071	2089	2107	3	6	9	12	15
12	0·2126	2144	2162	2180	2199	2217	2235	2254	2272	2290	3	6	9	12	15
13	0·2309	2327	2345	2364	2382	2401	2419	2438	2456	2475	3	6	9	12	15
14	0·2493	2512	2530	2549	2568	2586	2605	2623	2642	2661	3	6	9	12	16
15	0·2679	2698	2717	2736	2754	2773	2792	2811	2830	2849	3	6	9	13	16
16	0·2867	2886	2905	2924	2943	2962	2981	3000	3019	3038	3	6	9	13	16
17	0·3057	3076	3096	3115	3134	3153	3172	3191	3211	3230	3	6	10	13	16
18	0·3249	3269	3288	3307	3327	3346	3365	3385	3404	3424	3	6	10	13	16
19	0·3443	3463	3482	3502	3522	3541	3561	3581	3600	3620	3	7	10	13	16
20	0·3640	3659	3679	3699	3719	3739	3759	3779	3799	3819	3	7	10	13	17
21	0·3839	3859	3879	3899	3919	3939	3959	3979	4000	4020	3	7	10	13	17
22	0·4040	4061	4081	4101	4122	4142	4163	4183	4204	4224	3	7	10	14	17
23	0·4245	4265	4286	4307	4327	4348	4369	4390	4411	4431	3	7	10	14	17
24	0·4452	4473	4494	4515	4536	4557	4578	4599	4621	4642	4	7	11	14	18
25	0·4663	4684	4706	4727	4748	4770	4791	4813	4834	4856	4	7	11	14	18
26	0·4877	4899	4921	4942	4964	4986	5008	5029	5051	5073	4	7	11	15	18
27	0·5095	5117	5139	5161	5184	5206	5228	5250	5272	5295	4	7	11	15	18
28	0·5317	5340	5362	5384	5407	5430	5452	5475	5498	5520	4	8	11	15	19
29	0·5543	5566	5589	5612	5635	5658	5681	5704	5727	5750	4	8	12	15	19
30	0·5774	5797	5820	5844	5867	5890	5914	5938	5961	5985	4	8	12	16	20
31	0·6009	6032	6056	6080	6104	6128	6152	6176	6200	6224	4	8	12	16	20
32	0·6249	6273	6297	6322	6346	6371	6395	6420	6445	6469	4	8	12	16	20
33	0·6494	6519	6544	6569	6594	6619	6644	6669	6694	6720	4	8	13	17	21
34	0·6745	6771	6796	6822	6847	6873	6899	6924	6950	6976	4	9	13	17	21
35	0·7002	7028	7054	7080	7107	7133	7159	7186	7212	7239	4	9	13	18	22
36	0·7265	7292	7319	7346	7373	7400	7427	7454	7481	7508	5	9	14	18	23
37	0·7536	7563	7590	7618	7646	7673	7701	7729	7757	7785	5	9	14	18	23
38	0·7813	7841	7869	7898	7926	7954	7983	8012	8040	8069	5	9	14	19	24
39	0·8098	8127	8156	8185	8214	8243	8273	8302	8332	8361	5	10	15	20	24
40	0·8391	8421	8451	8481	8511	8541	8571	8601	8632	8662	5	10	15	20	25
41	0·8693	8724	8754	8785	8816	8847	8878	8910	8941	8972	5	10	16	21	26
42	0·9004	9036	9067	9099	9131	9163	9195	9228	9260	9293	5	11	16	21	27
43	0·9325	9358	9391	9424	9457	9490	9523	9556	9590	9623	6	11	17	22	28
44	0·9657	9691	9725	9759	9793	9827	9861	9896	9930	9965	6	11	17	23	29

NATURAL TANGENTS.

°	0'	6'	12'	18'	24'	30'	36'	42'	48'	54'	1'	2'	3'	4'	5'
45	1·0000	0035	0070	0105	0141	0176	0212	0247	0283	0319	6	12	18	24	30
46	1·0355	0392	0428	0464	0501	0538	0575	0612	0649	0686	6	12	18	25	31
47	1·0724	0761	0799	0837	0875	0913	0951	0990	1028	1067	6	13	19	25	32
48	1·1106	1145	1184	1224	1263	1303	1343	1383	1423	1463	7	13	20	26	33
49	1·1504	1544	1585	1626	1667	1708	1750	1792	1833	1875	7	14	21	28	34
50	1·1918	1960	2002	2045	2088	2131	2174	2218	2261	2305	7	14	22	29	36
51	1·2349	2393	2437	2482	2527	2572	2617	2662	2708	2753	8	15	23	30	38
52	1·2799	2846	2892	2938	2985	3032	3079	3127	3175	3222	8	16	24	31	39
53	1·3270	3319	3367	3416	3465	3514	3564	3613	3663	3713	8	16	25	33	41
54	1·3764	3814	3865	3916	3968	4019	4071	4124	4176	4229	9	17	26	34	43
55	1·4281	4335	4388	4442	4496	4550	4605	4659	4715	4770	9	18	27	36	45
56	1·4826	4882	4938	4994	5051	5108	5166	5224	5282	5340	10	19	29	38	48
57	1·5399	5458	5517	5577	5637	5697	5757	5818	5880	5941	10	20	30	40	50
58	1·6003	6066	6128	6191	6255	6319	6383	6447	6512	6577	11	21	32	43	53
59	1·6643	6709	6775	6842	6909	6977	7045	7113	7182	7251	11	23	34	45	56
60	1·7321	7391	7461	7532	7603	7675	7747	7820	7893	7966	12	24	36	48	60
61	1·8040	8115	8190	8265	8341	8418	8495	8572	8650	8728	13	26	38	51	64
62	1·8807	8887	8967	9047	9128	9210	9292	9375	9458	9542	14	27	41	55	68
63	1·9626	9711	9797	9883	9970	0057	0145	0233	0323	0413	15	29	44	58	73
64	2·0503	0594	0686	0778	0872	0965	1060	1155	1251	1348	16	31	47	63	78
65	2·1445	1543	1642	1742	1842	1943	2045	2148	2251	2355	17	34	51	68	85
66	2·2460	2566	2673	2781	2889	2998	3109	3220	3332	3445	18	37	55	73	91
67	2·3559	3673	3789	3906	4023	4142	4262	4383	4504	4627	20	40	60	79	99
68	2·4751	4876	5002	5129	5257	5386	5517	5649	5782	5916	22	43	65	87	108
69	2·6051	6187	6325	6464	6605	6746	6889	7034	7179	7326	24	47	71	95	119
70	2·7475	7625	7776	7929	8083	8239	8397	8556	8716	8878	26	52	78	104	130
71	2·9042	9208	9375	9544	9714	9887	0061	0237	0415	0595	29	58	87	116	144
72	3·0777	0961	1146	1334	1524	1716	1910	2106	2305	2506	32	64	97	129	161
73	3·2709	2914	3122	3332	3544	3759	3977	4197	4420	4646	36	72	108	144	180
74	3·4874	5105	5339	5576	5816	6059	6305	6554	6806	7062	41	81	122	163	203
75	3·7321	7583	7848	8118	8391	8667	8947	9232	9520	9812	46	93	139	186	232
76	4·0108	0408	0713	1022	1335	1653	1976	2303	2635	2972	53	107	160	214	267
77	4·3315	3662	4015	4373	4737	5107	5483	5864	6252	6646	62	124	186	248	310
78	4·7046	7453	7867	8288	8716	9152	9594	0045	0504	0970	73	146	220	293	366
79	5·1446	1929	2422	2924	3435	3955	4486	5026	5578	6140	87	175	263	350	438
80	5·671	5·730	5·789	5·850	5·912	5·976	6·041	6·107	6·174	6·243					
81	6·314	6·386	6·460	6·535	6·612	6·691	6·772	6·855	6·940	7·026					
82	7·115	7·207	7·300	7·396	7·495	7·596	7·700	7·806	7·916	8·028					
83	8·144	8·264	8·386	8·513	8·643	8·777	8·915	9·058	9·205	9·357					
84	9·51	9·68	9·84	10·02	10·20	10·39	10·58	10·78	10·99	11·20		Differences			
85	11·43	11·66	11·91	12·16	12·43	12·71	13·00	13·30	13·62	13·95		untrustworthy			
86	14·30	14·67	15·06	15·46	15·89	16·35	16·83	17·34	17·89	18·46		here.			
87	19·08	19·74	20·45	21·20	22·02	22·90	23·86	24·90	26·03	27·27					
88	28·64	30·14	31·82	33·69	35·80	38·19	40·92	44·07	47·74	52·08					
89	57·29	63·66	71·62	81·85	95·49	114·6	143·2	191·0	286·5	573·0					

The black type indicates that the integer changes.

°	Rad.	°	Rad.	°	Rad.	′	Rad.	′	Rad.
0	0·0000	30	0·5236	60	1·0472	0	·0000	30	·0087
1	0·0175	31	0·5411	61	1·0647	1	·0003	31	·0090
2	0·0349	32	0·5585	62	1·0821	2	·0006	32	·0093
3	0·0524	33	0·5760	63	1·0996	3	·0009	33	·0096
4	0·0698	34	0·5935	64	1·1170	4	·0012	34	·0099
5	0·0873	35	0·6109	65	1·1345	5	·0015	35	·0102
6	0·1047	36	0·6283	66	1·1519	6	·0017	36	·0105
7	0·1222	37	0·6458	67	1·1694	7	·0020	37	·0108
8	0·1396	38	0·6632	68	1·1868	8	·0023	38	·0111
9	0·1571	39	0·6807	69	1·2043	9	·0026	39	·0113
10	0·1745	40	0·6981	70	1·2217	10	·0029	40	·0116
11	0·1920	41	0·7156	71	1·2392	11	·0032	41	·0119
12	0·2094	42	0·7330	72	1·2566	12	·0035	42	·0122
13	0·2269	43	0·7505	73	1·2741	13	·0038	43	·0125
14	0·2443	44	0·7679	74	1·2915	14	·0041	44	·0128
15	0·2618	45	0·7854	75	1·3090	15	·0044	45	·0131
16	0·2793	46	0·8029	76	1·3265	16	·0047	46	·0134
17	0·2967	47	0·8203	77	1·3439	17	·0049	47	·0137
18	0·3142	48	0·8378	78	1·3614	18	·0052	48	·0140
19	0·3316	49	0·8552	79	1·3788	19	·0055	49	·0143
20	0·3491	50	0·8727	80	1·3963	20	·0058	50	·0145
21	0·3665	51	0·8901	81	1·4137	21	·0061	51	·0148
22	0·3840	52	0·9076	82	1·4312	22	·0064	52	·0151
23	0·4014	53	0·9250	83	1·4486	23	·0067	53	·0154
24	0·4189	54	0·9425	84	1·4661	24	·0070	54	·0157
25	0·4363	55	0·9599	85	1·4835	25	·0073	55	·0160
26	0·4538	56	0·9774	86	1·5010	26	·0076	56	·0163
27	0·4712	57	0·9948	87	1·5184	27	·0079	57	·0166
28	0·4887	58	1·0122	88	1·5359	28	·0081	58	·0169
29	0·5061	59	1·0297	89	1·5533	29	·0084	59	·0172
30	0·5236	60	1·0472	90	1·5708	30	·0087	60	·0175

Rad.	Degrees.
0·001	0·06
0·002	0·11
0·003	0·17
0·004	0·23
0·005	0·29
0·006	0·34
0·007	0·40
0·008	0·46
0·009	0·52
0·01	0·57
0·02	1·15
0·03	1·72
0·04	2·29
0·05	2·86
0·06	3·44
0·07	4·01
0·08	4·58
0·09	5·16
0·1	5·73
0·2	11·46
0·3	17·19
0·4	22·92
0·5	28·65
0·6	34·38
0·7	40·11
0·8	45·84
0·9	51·57
1	57·30
2	114·59
3	171·89
4	229·18
5	286·48
6	343·77

WEIGHTS AND MEASURES.

1. Measure of Distance: or Linear Measure.

12 inches = 1 foot.	10 millimetres = 1 centimetre.	1 cm. = 0·394 in.
3 feet = 1 yard.	100 centimetres = 1 metre.	1 m. = 1·09 yds.
1760 yards = 1 mile.	1000 metres = 1 kilometre.	= 39·4 in.
100 links = 1 chain.		1 km. = 0·621 mi
= 22 yards.		1 in. = 2·54 cm.

Less important:

1 pole or rod	= 5·5 yards.	
1 furlong	= 220 yards.	10 decimetres = 1 metre.
1 cable	= 608 feet.	
1 sea mile	= 6080 feet.	

2. Measure of Area : or Square Measure.

144 sq. inches = 1 sq. foot.	100 sq. cm. = 1 sq. dm.	1 sq. cm. = 0·155 sq. in.
9 sq. feet = 1 sq. yard.	100 sq. dm. = 1 sq. m.	1 hectare = 2·47 acres.
484 sq. yards = 1 sq. chain.	10,000 sq. m. = 1 hectare.	1 sq. in. = 6·45 sq. cm.
10 sq. chains = 1 acre.	100 hects. = 1 sq. km.	
640 acres = 1 sq. mile.		

1 sq. rod or pole = 30·25 sq. yards.   1 are = 100 sq. metres.
1 rood = 40 sq. poles.
4 roods = 1 acre.

3. Measure of Volume : or Cubic Measure, Measure of Capacity, and Liquid Measure.

1728 cu. in. = 1 cu. ft.	1000 c. c. = 1 cu. dm.	1 c. c. = 0·0610 cu. in.
27 cu. ft. = 1 cu. yd.	= 1 litre.	1 litre = 0·220 gal.
1 gallon = 277 cu. in.		1 cu. in. = 16·4 c. c.
1 cu. ft. = 6·24 gallons.		

2 pints = 1 quart.   100 centilitres = 1 litre.
4 quarts = 1 gallon.   100 litres = 1 hectolitre.
8 gallons = 1 bushel.
8 bushels = 1 quarter.

4. Measure of Weight.

16 ounces = 1 pound.	1000 grms. = 1 kg.	1 kg. = 2·20 lbs.
112 pounds = 1 cwt.	1000 kgs. = 1 metric tonne.	1 tonne = 0·984 ton.
20 cwt. = 1 ton.		1 lb. = 454 grms.

7000 grains = 1 pound.   28 pounds = 1 quarter.
14 pounds = 1 stone.   100 pounds = 1 cental.

CONSTANTS.

$\pi = 3·1416$ $\log_{10}\pi = 0·4971$

1 radian = 57·296 degrees.

$e = 2·7183$ $\log_{10}e = 0·4343$

$\log_e N = 2·3026 \log_{10} N$; $\log_{10} N = 0·4343 \log_e N$.

Earth's mean radius = 3960 miles = 6·371 × 10⁸ cm.

A velocity of 60 miles per hour = 88 feet per second.

A velocity of 1 knot = 1 sea-mile per hour = 1·7 feet per sec. (nearly).

$g = 32·2$ ft. per sec. per sec. or 981 cm. per sec. per sec.

Length of seconds pendulum (Greenwich) = 39·139 in. = 99·413 cm.

1 atmosphere = 760 mm. or 29·9 in. of mercury = 1·03 kg. per sq. cm. = 14·7 lbs. per sq. in.

Velocity of sound in air is about 1100 ft. per sec. = 3·3 × 10⁴ cm. per sec.

Velocity of light *in vacuo* = 186,300 miles per sec. = 3 × 10¹⁰ cm. per sec.

ANSWERS TO EXERCISES

EXERCISES I.

1. 8·7 lbs.
2. 99 lbs.   4. 72 lbs.   5. 13 lbs.
6. Extracting component 8·66 lbs.; bending component 5 lbs.
7. 78 lbs. in the short wire and 26 lbs. in the long one.
8. Pressure 98·5 lbs.; sliding force 17·4 lbs.
9. Between 11° and 12°.   10. ·33 ton.
11. 46·7 lbs.   12. 47 lbs. and 63 lbs.   13. 223 lbs.
14. Resultant = 41·4 lbs. and acts along the line bisecting $\angle P_1 O P_2$ in the opposite direction from OP_3.   15. 25·8 lbs.

EXERCISES II,

1. 16·4 lbs.   2. $W = 110·77$ lbs. Pressure on fulcrum = 614·77 lbs.
3. 11·05 ins. from end weighted with 35 lbs.
4. $3\frac{9}{17}$ ft. from end weighted with 36 lbs.   5. 41·8 ins.
6. 64·5 lbs. per sq. in. above atmosphere.   7. 12·5 lbs.
8. 17·7 lbs.   9. ·866 ton.
10. 3·83 ins. nearly from the force of 10 lbs.   11. 69·2 lbs.
12. 3·55 tons, 2·45 tons.   13. 17,000 lbs.
14. Halfway along a line joining the apex to one-third of the base.
15. 160 lbs.   16. Loses $\dfrac{(a-b)^2}{2ab}$ lb. per lb.   17. 3 cwt.

EXERCISES III.

1. 1,080,000 ft.-lbs.   2. 1,752,000 ft.-lbs.   3. 127·4 H.P.
4. $22\frac{8}{11}$ H.P.   5. 352 cub. ft.   6. 42,240,000 ft.-lbs.
7. 27,456 ft.-lbs.   8. $68\frac{2}{11}$ H.P.   9. 450 H.P.
10. 5,430,000 ft.-lbs.   11. ·29 H.P.   12. 119,000,000 ft.-lbs.
13. 179·2 H.P.; 224 H.P.   14. 218 H.P.   15. 2,704,000 ft.-lbs.
16. 720 tons.   17. 247,000 ft.-lbs.; 112,700 ft.-lbs.; 134,300 ft.-lbs.

EXERCISES IV.

1. 195·45 lbs.   2. 99·77% efficiency.
3. Required pull, $166\frac{2}{3}$ lbs.; Efficiency = 83·33%.
                Mechanical advantage = 3, Velocity ratio = 3·6 : 1.
4. 132·2 H.P.
5. Necessary force parallel to plane = $1\frac{1}{5}$ tons wt.
    Necessary force parallel to base = $1\frac{1}{2}$ tons wt.
6. 6·125 lbs.   7. 12,600 in.-lbs.   8. 57·14 H.P.
9. $V_r = 326·9$; $W = 1552$ lbs; $\eta = 8·5\%$.   10. 22·3 H.P.
11. $7\frac{1}{2}$ H.P.   12. 99·7 H.P.   13. 34·3 H.P.
14. 15·7 H.P   15. 91%.   16. 26%.

ANSWERS TO EXERCISES

EXERCISES V.

1. 40·1 ft. per sec.   **2.** 55·04 miles per hr.   **3.** 100·6 ft.
4. Heights fallen, 9·82 ft.;   0·9676 ft.;   0·0966 ft.;   Average velocities, 98·21 ft. per sec.; 96·76 ft. per sec.; 96·616 ft. per sec.
5. 1·47 ft. per sec.²   **6.** 80 miles per hour after ·2 hour.
7. 7·56 miles.   **8.** 3657 cms. per min.
9. 5·87 ft. per sec. per sec.   **10.** 75 miles per hr.   **11.** 100 yds.
12. 22 ft. per sec. per sec.   **13.** 1370 yds.   **14.** 60 miles per hour.

EXERCISES VI.

1. 55 feet per sec. at 37° to the direction of the train's motion.
2. 43 ft. per sec.   **3.** 10 miles an hour from the N.W.
4. 46·2 ft. per sec.; 83·25° to the circumference.   **5.** 27·3 secs.
6. 5 secs.   **7.** $\frac{10}{11}$ min.; 1¼ mins.; 20$\frac{5}{11}$ secs.   **8.** 35·5 ft. per sec.
9. 8·03 miles.   **10.** 25° 37′ N. of E.; 3·89 miles per hour.

EXERCISES VII.

1. 27 ft.-lbs.   **2.** 1880 yds.   **3.** 8430 ft. lbs.
4. 6170 lbs.   **5.** Energy = 93,000 ft.-lbs.; Pressure exerted = 18,600 lbs.
6. 3 lbs.   **7.** 220 ft. tons; ·8 ft. per sec. increase in velocity.
8. 25·5 ft. per sec.; 7700 ft.   **9.** 3110 lb. sec.
10. 1560 lbs.; 8·8 ft.   **11.** 760 feet.   **12.** 2880 feet.
13. 45·6 tons.   **14.** 418 lbs.
15. ·64 ft. per sec.²; 38·4 ft. per sec.   **16.** 2904 lbs.

EXERCISES VIII.

1. 3880 lbs.   **2.** 5·24 tons.   **3.** 845 lbs.; 1000 lbs.; 1155 lbs.
4. 221 lbs.   **5.** 12·9 lbs.   **6.** 58·8 ft. per sec.; more.
7. 128 ft. per sec.   **8.** 8·8 ft. per sec.; come to rest.

EXERCISES IX.

1. ·0006.   **2.** ·1 ft.   **3.** − ·0005.
4. Stress 16,000 lbs. per sq. in.; strain ·000625: E = 25,600,000.
5. ·0256 ft. = ⅓ in. nearly.   **6.** 12,000 in.-lbs.; 345 in.-lbs.
7. 3410 lbs. per sq. in.   **8.** 2·026 ins.   **9.** 14·02 tons.
10. 48,600 lbs. per sq. in.
11. Modulus of elasticity = 22,500,000 lbs. per sq. in.
12. Work done = 6 in. lbs.   **13.** 2700 ft.   **14.** 6 in. lbs.
15. 5·45 tons per sq. in.   **16.** 3¼ ins. diameter.
17. ·00074 in. longitudinal; ·000185 in. transverse.   **18.** 2180 lbs.

EXERCISES X.

1. 30 tons.   **2.** ⅞ in., double row at 4-inch pitch; 56%.   **3.** ·57.
4. Resistance to shearing of rivets = 18·85 tons; resistance to tearing of plates = 18 tons; thickness of cover plates should be $\frac{5}{16}$ in.; efficiency = 66·7 %.
5. 5 rivets; yes.   **6.** 120 tons.   **7.** 5½ ins.   **9.** 2·23 ins. nearly.
10. 3·71 ins.   **11.** 48·6 lbs. per sq. in. about.   **12.** ·357 in.

EXERCISES XI.

1. $AB = 7$; $BC = -7.7$; $CA = +15$; $BD = -13$ tons.
2. $AB = A'B' = +4.62$; $BD = B'D = -4.62$; $AD = A'D = -2.30$;
   $BB' = +2.30$ tons.
3. Top bars, $+48.4$, 62.2, 41.4, 20.5 tons.
   Bottom bars, -24.6, 55.6, 51.8, 30.8, 10.1 tons.
   Diagonals, ± 48.4, ± 13.8, remainder ± 20.9 tons.
4. $BC = +3.5$; $CD = +2.0$; $AE = -5.59$; $EF = -2.98$; $FD = -2.24$;
   $\qquad\qquad\qquad BE = +2.12$; $EC = -1.86$; $CF = +9.4$ tons.
5. Force in $DD' = 6.5$ tons.   6. 2.69 tons.
7. -10 tons in stay; $+8.8$ tons in each leg.

EXERCISES XII.

1. B.M.'s 1410, 870 and 480 lbs. ft.
   S.F.'s 180, 180 and 80 lbs.
2. B.M.'s, 0, 1400, 2600, 3600, 4400, 5000, 5400, 5600, 5625 (centre) tons ft.
   S.F.'s. 150, 130, 110, 90, 70, 50, 30, 10, 0 (middle).
3. B.M. (at centre) = 62,500 lbs. ft.
   S.F. (at centre) = 0.
   B.M. (15 ft. from end) = 52,500 lbs. ft.
   S.F. (15 ft. from end) = 2000 lbs.
4. B.M. (at fixed end)   = 91,875 lbs. ft.
   B.M. (15 ft. from fixed end) = 30,000 lbs. ft.
   B.M. (25 ft. from fixed end) = 7500 lbs. ft.
   S.F. (at fixed end)   = 5250 lbs.
   S.F. (15 ft. from fixed end) = 3000 lbs.
   S.F. (25 ft. from fixed end) = 1500 lbs.
5. (At fixed end)   B.M. = 105,000 lbs. ft., S.F. = 6000 lbs.
   (15 ft. from fixed end) B.M. = 31,875 lbs. ft., S.F. = 3750 lbs.
   (25 ft. from fixed end) B.M. = 7500 lbs. ft.,   S.F. = 1500 lbs.
6. (At centre) B.M. = 45 tons ft., S.F. = 2 tons.
   (5 ft. from end nearest which is wt.) B.M. = 20 tons ft., S.F. = 4 tons.
   (5 ft. from other end) B.M. = 10 tons ft., S.F. = 2 tons.
7. Reaction at support $A = \frac{3}{7}\frac{3}{7}$ tons.
   Reaction at support $B = \frac{0}{7}\frac{0}{7}$ tons.
   At C, B.M. = $44\frac{1}{2}$ tons ft., S.B. = $2\frac{4}{7}$ tons.
   At D, B.M. = 73 tons ft.,   S.F. = $2\frac{4}{7}$ tons.
   At E, B.M. = 70 tons ft.,   S.F. = $4\frac{3}{7}$ tons.
8. If A is one end of the axle and B, C, D points distance 4 ins. apart:
   Then B.M. at $A = 0$; at B, C, D, etc. = 20 tons ins.
   S.F. at $A = 2\frac{1}{2}$; at $B = 2\frac{1}{2}$; from B to other end = 0.
9. At point 5 ft. from one end B.M. = 6900 lbs. ft.,   S.F. = 930 lbs.
   At point 8 ft. from one end B.M. = 20,400 lbs. ft., S.F. = 390 lbs.
10. B.M. = 256,250 lbs. ft.; S.F. = 4000 lbs.
11. 96 lbs.   12. B.M. = 120 tons ft.; S.F. = 1 ton.
13. At centre   B.M. = 648 lbs. ft.,   S.F. = 0.
    At 1st quarter B.M. = 486 lbs. ft.,   S.F. = 72 lbs.
    At end   B.M. = 0,   S.F. = 144 lbs.

ANSWERS TO EXERCISES 313

EXERCISES XIII.

1. 2 ins. from the base.
2. 2·5 ins. from the base.
3. $\frac{1}{10}$ of its length from the centre.
4. 1·122 ft. from the point of contact.
5. 2·33 ins. from the centre of the rod.
6. 24·16 ins. from bottom.
7. ·6R from centre of circle.
8. 18,900 lbs.
9. Height = base × 1·732.
10. ·831 in. from base.
11. Centre of gravity = 3·747 ins. nearly from lower flange XY.
12. Centre of gravity is a point ·83 in. nearly from the 4-in. side and 1·33 ins. nearly from the 3-in. side.
13. 2·6 ins.
14. A point 2·75 ft. from the end weighted with 2 lbs.
15. $\frac{1}{10}$ ft.
16. $\frac{9l}{16}$ from right-hand end.
17. 4⅞ ins. from base.
18. 2·97 ft.

EXERCISES XIV.

1. 0·179.
2. 10 lbs.
3. ·15.
4. (i) 371·7 lbs. (ii) 370·7 lbs. (iii) 370·0 lbs.
5. 10 tons at 30% to horizon.
6. 60·5 ft.
8. 4·6.
9. 9 ins.
10. 8°; 800 sq. ft.
11. 91 lbs.; 28%.
12. 50 ft.

EXERCISES XV.

1. 2,744,000 ft.-lbs.
2. 2·57 lbs.
3. 47·5 revs. per min.
4. 1410 lbs.
5. 28·6 revs. per min.
6. 15 miles per hr.
7. Centrifugal force = 8 tons approx.; 38·8 tons at 12° to vert.
8. 42·6 miles per hr.
9. 17,000 ft.-lbs.
10. 38·2 revs. per min.
11. 80,000 ft.
12. 6000 yds.

EXERCISES XVI.

1. (a) 84°; (b) ·55 of stroke from the back dead point.
2. (1) 12,403 lbs.; (2) 3100·75 lbs.
4. 640 ft. per min.
6. 1·28:1; 4 ins.

EXERCISES XVII.

1. 3 ft. 10 ins. nearly.
2. 18.
3. Difference in tension = 583·3 lbs.; tension on sides = 291·7 lbs. and 875 lbs.
4. 3 tons.
5. 250 lbs.
6. 675 lbs.
7. 37·3 lbs.
8. 912 revs. per min.
9. ·91 H.P.
10. 13·37 lbs.
11. 5½ ins.
12. 7½ ins.
13. 55·8 ft.
14. $\frac{83}{20} \times \frac{78}{30}$.
15. 455, 23·8.
16. 45·3%: 27·2 H.P.

INDEX

Acceleration 92–100
  relation to force 117
Angle of friction 225
  of repose 233
Angular velocity 243

Back gear for lathes 287
Balancing rotating parts 250
Beams and girders 188–200
  bending moment 189
  reactions 21
  shearing force 189
Belt gearing 273–279
Bending moments 189–200
Bevel gearing 281
Bicycle two-speed gear 73
Bow's notation 3
Brake horse-power 78

Cams 267
Cantilevers 190–193
Cast iron, stress-strain diagram of 143
Cement and concrete 144
Centrifugal and centripetal force 245
Centroid and centre of gravity 203–220
Clerk Maxwell diagrams 178
Columns 156
Compressive stress 140
Cone, centre of gravity of 212
Conservation of energy 40
Counterbracing 175
Couples 31
Crank and connecting-rod mechanism 258
Crow-bar 53
Curved path, motion in 242–256
Cycloid curve 104
Cylinders, strength of 169

Deficient frames 174
Diametral pitch 280
Differential pulley block 67

Efficiency
  of machines 55
  of riveted joints 168

Effort curve 42
  mean 49
Elastic bodies 139
  limit 141
Energy
  conservation of 40
  definition and kinds 39
  useful 41
Equilibrant 8
Equilibrium 8
  kinds of 221
  under three forces 25
Experiments
  bicycle two-speed gear 7
  centre of gravity and centroid 220
  errors of 13
  friction 238
  inclined plane 60
  moments 18
  polygon of forces 12
  reaction of jet 133
  roof-truss 184
  triangle of forces 4
  Weston pulley block 72
  wire, strength of 147

Factor of safety 151
Force
  polygon 11
  triangle of 3
  unit of 2
  see also effort, reaction, resistance, etc., and various kinds of machines and structures
Framed structures 174–185
Friction 224–240
  angle of 225
  rolling 228
  static and kinetic 224

Gearing 273–290
Gear trains 286
Governor 248
Gravity acceleration 97
  centre of 203–220

Hodograph 242
Hooke's law 139

Horse-power
  brake 78
  definition 38
  indicated 77

Idle gear wheels 285
Impact and impulse 129
Inclined plane 57–63, 229–233
Indicated horse-power 77
Instantaneous centre 259
Involute gearing 280

Kinetic energy 40, 113
  friction 224

Lathe back gear 287
  headstock gear 276
  lead screw drive 289
Leverage 16
Lever safety valve 23
Limit, elastic 141
Line of pressure 11
Link and vector polygon construction
  27, 32
Lubrication, see Friction

Machines 52–81
  actual performance 70
  reversing 66
Mean effort 49
Mechanical advantage 53
Mechanisms 258–270
Method of moments or sections 183
Module of toothed gearing 280
Modulus, elastic 145
Moments
  of forces 16–33
  method of, for frames 183
Momentum 119–137
Motor tracks 247

Newton's laws of motion 1, 21, 117,
  124

Parabola, centroid of 219
Pawl and ratchet mechanism 270
Pile driver 135
Pillars 156
Pipes, strength of 169
Pitch circle 280
Planing machine, belt drive 278
Poisson's ratio 146
Polygon of forces 11
Potential energy 40
Power (see also Horse-power) 38
Projectiles 252–256
Pulley tackle 67–70
Pyramid, centre of gravity of 211

Quadrilateral, centroid of 216
Quick-return mechanism 263

Rack and pinion 281
Railway curves 247
Ratchet mechanism 270
Reaction
  Newton's law 124
  of beams 21
Reciprocal figures 178
Recoil of guns 133
Redundant frames 175
Relative velocity 103–111
Repose, angle of 233
Resilience 152
Resistances 45
Resultant, definitions of 3
Reversing gear train 289, 290
Reversing machines 66
Rigidity modulus 145
Ritter's method for frames 185
Riveted joints 161–169
Rolling friction 228
Roof-truss 177–185
Rotating bodies 75

Safety valve 23
Scalar quantities 1
Screw
  jack 65
  with friction 233
  without friction 63
Sections, method of, for frames 183
Semicircle, centroid, etc. of 220
Shear
  in beams 189–200
  legs 184
  modulus 145
  stress 140
Ships, relative velocities of 108
Space curve 85
Speed, see Velocity
Speed-cones 276
Spiral gearing 282
Stability of wall 22
Steam-engine foundation, thrust on
  5
Steel, stress-strain, diagram of 142
Strain 139
Stress
  definition 139
  dynamic 153
  -strain diagrams 141–150
  temperature 156
  working 151
Struts 156

Tensile stress 140
Timber, strength of 144
Toggle mechanism 265
Toothed gearing 279–290
Torque 76
Trapezium, centroid of 212
Triangle, etc., centroid of 210

Triangle of forces 3
Tripods 184

Useful energy 41

Vector polygon construction 10
Vector quantities 1
Velocity
   angular 243
   curves 87
   definitions 83
   in mechanisms 259
   ratio in gearing 275, 279, 283
   ratio of machines 56
   relative 103–111
   uniform 83

Velocity
   variable 84
Virtual centre 260

Warren girder 180
Watt's parallel motion 262
Weight as unit of force 1
Weston's pulley block 68
Work 36
   resistance, against 44
   variable force done by 42
Working stress 151
Worm gearing 282

Yield point 141
Young's modulus 145

Printed in the United States
By Bookmasters